THE OPEN UNIVERSITY

Science:
A Second Level Course

S299
GENETICS

Prepared by a Course Team for the Open University

THE OPEN UNIVERSITY PRESS

Course Team

Chairman and General Editor
Steven Rose

Unit Authors
Norman Cohen (*The Open University*)
Terence Crawford-Sidebotham (*University of York*)*
Denis Gartside (*University of Hull*)
David Jones (*University of Hull*)
Steven Rose (*The Open University*)
Derek Smith (*University of Birmingham*)
Mike Tribe (*University of Sussex*)
Robert Whittle (*University of Sussex*)

*Consultant

Editor
Jacqueline Stewart

Other Members
Bob Cordell (*Staff Tutor*)
Mae-Wan Ho*
Jean Holley (*Technician*)
Stephen Hurry
Roger Jones (*BBC*)
Aileen Llewellyn (*Course Assistant*)
Michael MacDonald-Ross (*IET*)
Jean Nunn (*BBC*)
Pat O'Callaghan (*Evaluation*)
Jim Stevenson (*BBC*)

* From January 1976

The development of this Course was supported by a grant from
the Nuffield Foundation.

Maintenance Course Team

Chairman and General Editor
Mae-Wan Ho

Unit Authors (*2nd ed. Unit 6*)
Mae-Wan Ho
Pat O'Callaghan

Editor
Jacqueline Stewart

Other Members
Mary Bell (*Staff Tutor*)
Robin Harding (*Staff Tutor*)
Jean Holley (*Technician*)
Stephen Hurry
Roger Jones (*BBC*)
Aileen Llewellyn (*Course Assistant*)
Pat O'Callaghan (*Evaluation*)

The Open University Press,
Walton Hall, Milton Keynes.

First published 1977

Designed by the Media Development Group of the Open University.

Set by Composition House Ltd, Salisbury, Wiltshire.

Printed in Great Britain by Eyre and Spottiswoode Limited,
at Grosvenor Press, Portsmouth.

ISBN 0 335 04291 0

This text forms part of an Open University Course. The complete list of Units in the Course
appears at the end of this text.

For general availability of supporting material referred to in this text please write to the
Director of Marketing, The Open University, 12 Cofferidge Close, Stony Stratford, Milton
Keynes, MK11 1BY.

Further information on Open University Courses may be obtained from the Admissions
Office, The Open University, P.O. Box 48, Walton Hall, Milton Keynes, MK7 6AB.

2.1

6 Molecular Genetics

Contents

List of scientific terms used in Unit 6

Introduced in S100*	Developed in this Unit	Page No.	Developed in this Unit	Page No.
codon	ambivalent mutant	277	non-permissive strain	277
polynucleotide	base deletion	270	nonsense codon	273
transcription	base insertion	270	nonsense suppression	278
translation	base substitution	266	one enzyme–one gene hypothesis	253
	codon assignment	271	one polypeptide– one gene hypothesis	255
	colinearity	273	operator gene	283
	constitutive mutant	284	operon	283
	coordinate expression of genes	283	permissive strain	277
	degeneracy	268	point mutant	258
	deletion map	258	polar mutant	283
	deletion mutant	258	positive control	286a
	effector	283	promoter gene	283
	frameshift mutant	270	pseudo wild-type	269
	hot-spots	262	reading frame	269
	inducer	281	regulator gene	283
	intragenic complementation	255	repressor	283
	missense mutation	275	structural gene	283
	mispairing	265	suppression	269
	mutagens	266	transition	266
	negative control	286a	transversion	266
	non-inducible mutant	284		

* The Open University (1971) S100 *Science: A Foundation Course*, The Open University Press.

Objectives for Unit 6

After studying this Unit you should be able to:

1 Define, recognize the best definition of, and place in the correct context, the items in the list of scientific terms opposite.
(SAQ 3)

2 Explain the relationship between genes and enzymes in determining biochemical characteristics.
(ITQs 1 and 2; SAQ 1)

3 Analyse and interpret data from fine-structure genetic analysis.
(ITQs 3 and 4; SAQs 2–4)

4 Outline the molecular basis of mutation as revealed by studies on chemical mutagens.
(ITQs 5 and 6; SAQ 5)

5 Define the genetic code.
(ITQ 7)

6 Recall and interpret the results of genetic and biochemical experiments that helped to crack the genetic code.
(ITQs 7 and 8; SAQ 6)

7 Explain how the genetic code is read and punctuated.
(SAQs 7 and 8)

8 Explain the characteristics of enzyme induction and repression.
(ITQ 9)

9 Define the components of regulatory systems for the control of gene expression.
(ITQs 10, 11, 13 and 14)

10 Recall and interpret the genetic evidence for the existence of the components of the *lac* operon.
(ITQs 9, 10 and 12)

11 Explain the molecular basis of the gene.
(SAQs 9 and 10)

Study guide for Unit 6

Unit 6 gives an account of the synthesis of formal genetics and biochemistry; this has revolutionized both subjects over the last two decades. We had two main problems in writing this Unit: to confine ourselves to what is recognizably genetics rather than biochemistry, and to keep our discussion of one of the most exciting chapters of modern biology within the confines of a single Unit.

Our approach is predominantly an historical one, with the benefit of hindsight. Relevant recent findings will be mentioned insofar as they contribute to the understanding of each topic. Emphasis is placed on the logical development of concepts and ideas, therefore it is important to read the Unit Sections in sequence. All Sections are essential reading, but if you are *very* short of time, you may leave out Section 6.3 on mutagenesis and Section 6.5.5 on the control of gene expression in higher organisms.

Introduction to Unit 6

This Unit tells the story of the search for the molecular identity of the 'gene'. As first conceived by Mendel, the gene was a hereditary factor which could be passed on unchanged from generation to generation. For a long time, however, the physical basis of the factor of heredity remained unknown whereas the idea of 'gene' became very elaborate, and appeared in different guises to geneticists working in different areas.

Let us review the various concepts of the gene that we have encountered so far. In Unit 1, you saw that the early Mendelians regarded the gene as a unit of character: each gene was responsible for the development of a phenotypic character such as the colour of flowers or the shape of seeds. Thus, a gene could be recognized only if the character observed existed in alternative forms, or alleles (e.g. *wrinkled* versus *round* seeds), which showed segregation in genetic crosses.

Another level of sophistication was reached when geneticists realized that the one *character*–one gene idea was wrong. For example, in Unit 2, you saw that in *D. melanogaster* some mutant strains with brown eyes did not produce brown-eyed offspring when they were inter-crossed; instead, the F_1 flies had normal dull-red eyes. The mutant strains were said to complement one another in heterozygous combination, thereby restoring the normal phenotype. In order to explain the phenomenon of complementation, it was necessary to postulate that several gene functions (say, in the production of normal eye pigment) were involved in the expression of a normal character, and that each gene function was specified by one gene. Therefore, whenever complementation occurs between two mutant strains, they are generally considered to be deficient in *different* genes (and thus lacking in different gene functions). Conversely, the absence of complementation is taken to indicate that the two mutant strains are deficient in the same gene. This one *function*–one gene concept is fundamental, and complementation tests became the first step in any genetic analysis.

Side by side with the development of the function concept of the gene was the development of the structure concept. In Units 2 and 3, you saw how genes were assigned to chromosomes, and how results from genetic crosses were used to construct linkage maps representing the linear order of genes on chromosomes and their physical distances apart. The picture of a chromosome was that of a string of beads, each bead representing an individual gene. It had been known for some time that each gene, as defined by its locus on the chromosome, could exist in several mutant forms, or alleles, which did not complement one another in heterozygous combination. It was assumed that no recombination could take place between mutants in the same gene, as recombination was thought to occur only *between* genes. This idea proved to be quite erroneous: mutants that did not complement gave wild-type recombinants, but only at very low frequencies.

What then, are the exact relationships between the units of complementation, mutation and recombination? This problem exercised geneticists for a long time until the molecular basis of inheritance was finally elucidated in the 1960s. The first crucial steps were taken more than 20 years before, by Beadle and Tatum (1941), who studied biochemical defects in *Neurospora crassa*, and by Avery, MacLeod and McCarty (1944), who demonstrated clearly that DNA was the genetic material. The major developments in this exciting period will be described in this Unit, which will also explain the interrelationships of all the different concepts of the gene in molecular terms. In the last Section of the Unit we shall discuss the problem of the control of gene expression, which is ultimately involved in the translation of genes into phenotype.

The plan of the Unit is as follows.

Section 6.1 defines the gene biochemically as a unit of function which specifies a polypeptide.

Section 6.2 investigates the molecular dimensions of the units of complementation, mutation and recombination, and their interrelationships.

Section 6.3 shows how mutagenesis throws light on the molecular basis of mutation.

Section 6.4 relates the sequence of bases in DNA to the sequence of amino acids in the polypeptides.

Section 6.5 deals with the molecular basis of the control of gene expression.

6.1 The function of genes

As long ago as 1902, Garrod recognized that 'inborn errors of metabolism' in man were metabolic blocks due to defects in single genes (see *HIST**, Section H.5.1). One of the main functions of genes was, therefore, to control specific chemical reactions in metabolism. Beadle and Tatum, aware of the complexity of trying to identify the chemical basis of phenotypic changes, set out to determine how genes controlled biochemical reactions. Their pioneering work was done in the early and mid-1940s using the bread mould *Neurospora crassa*.

Beadle and Tatum began by selecting for auxotrophs (see Unit 2, Section 2.11) after irradiating fungal spores with X-rays to increase the mutation rate; they then analysed the auxotrophs in order to identify the sites of metabolic blocks. Early studies were also carried out by Srb and Horowitz on several mutant strains that required the amino acid arginine for growth. All of the mutants grew on minimal medium supplemented with arginine; two related amino acids, citrulline and ornithine, supported the growth of some mutants but not others. These results are summarized in Table 1.

Table 1 The growth response of arginine auxotrophs of *N. crassa*

Mutant	unsupplemented minimal medium	arginine	minimal medium plus citrulline	ornithine
A	−	+	+	+
B	−	+	+	−
C	−	+	−	−

(+ = growth; − = no growth)

QUESTION What can you infer from the data about the metabolic defect of each mutant and the sequence of reactions involving the metabolites? (The ability to grow on either citrulline or ornithine means that the particular mutant can convert these substances into arginine.)

ANSWER Mutant A can convert both ornithine and citrulline to arginine, and so it grows on all three supplemented media.

Mutant B can only convert citrulline to arginine; hence it does not grow on ornithine.

Mutant C cannot convert either citrulline or ornithine to arginine; hence there is no growth except on arginine.

The sequence of the biosynthetic reactions involved must be as shown in the following pathway (the metabolic block in each mutant is indicated):

```
          ────→ ornithine ────→ citrulline ────→ arginine
               ↑               ↑               ↑
  mutants      A               B               C
```

From the chemical structure of the intermediates, it appeared likely that each step in the biosynthetic pathway could be catalysed by a single enzyme. Considerations such as these gave rise to the *one enzyme–one gene hypothesis* which stated that each genetically distinct metabolic block was the result of the deficiency of a single enzyme caused by a mutation in one gene.

one enzyme–one gene hypothesis

Now look again at the pathway for the biosynthesis of arginine, taking note of the site of the metabolic block in each mutant. If the mutants are grown in minimal medium containing just enough arginine to be completely used up by growth, the mutants will try to synthesize their own arginine, but can take the synthesis only as far as their particular metabolic block.

* The Open University (1976) S299 HIST, *The History and Social Relations of Genetics*, The Open University Press. This text is to be studied in parallel with the Units of the Course.

QUESTION What substances would you expect to be accumulated in mutants A, B and C respectively, assuming that the biosynthetic pathway is as depicted above?

ANSWER Each mutant would accumulate the metabolite immediately *before* its particular block, that is, citrulline in C, ornithine in B, and in A, the immediate metabolic precursor of ornithine.

This simple principle has been used to work out the steps involved in complex biochemical pathways.

ITQ 1 Three auxotrophic strains of *N. crassa* (1–3) requiring substance X were grown in minimal medium containing just enough X to support growth. Each strain was harvested and homogenized to give an extract. The individual extracts were tested for their ability to support the growth of each mutant strain on minimal medium supplemented with the extract. The results were as follows:

| | Growth response of mutants | | |
Extract	1	2	3
1	−	−	−
2	+	−	−
3	+	+	−

+ = growth; − = no growth)

Chemical analyses were performed on the individual extracts and the metabolites accumulated in each mutant were as follows:

Mutant	Metabolite accumulated
1	W
2	Y
3	Z

From the data given, reconstruct the metabolic pathway leading up to X and indicate the site of the metabolic block in each mutant.

The answers to the ITQs are on pp. 286j–l.

The one enzyme–one gene hypothesis gave a biochemical definition of the gene— that which specifies a particular enzyme. This means that we can draw inferences about the existence of a gene from the biochemical identification of an enzyme, *without performing a genetic cross.* The one enzyme–one gene hypothesis has received considerable support from the discovery of many hereditary diseases in humans which involve enzyme deficiencies.

6.1.1 Genes and proteins

In fact, enzymes are just one class of proteins (S100[1]) all of which are specified by genes. Proteins include, besides the enzymes that catalyse biochemical reactions, structural proteins that form the matrixes of tissues or cells (e.g. collagen in skin) and transport proteins that carry essential molecules to all parts of the body (e.g. haemoglobin transports oxygen). Although the protein–gene relationship is quite general, it is not always in the ratio 1 : 1. Let us see how the study of protein structure has altered our ideas concerning the protein–gene relationship.

Proteins are composed of one or more polypeptides or subunits; each polypeptide is made up of a chain of amino acids joined end-to-end by peptide bonds (see Section 6.4.5). As each polypeptide is synthesized in the cell, it becomes coiled and folded in a particular way, depending on its amino-acid sequence, to form a three-dimensional structure. It is this three-dimensional structure that makes up the subunit of a protein. A functional protein, therefore, may consist of a single polypeptide subunit, or it may consist of two or more subunits. The shapes of the subunits in a given protein are such that they can fit together or associate in a specific way (somewhat like the pieces of a three-dimensional jigsaw puzzle) to ensure the stability and functioning of the resultant protein molecule. Some proteins are made up of subunits that are all the same whereas other proteins contain different kinds of polypeptides. Where different polypeptides are involved each polypeptide is specified by a different gene. This *one polypeptide–one gene* relationship was actually discovered through the analysis of the chemical structure of normal and mutant haemoglobins. The haemoglobin molecule consists of two α- and two β-polypeptide subunits (see Fig. 1).

one polypeptide–one gene hypothesis

Sickle-cell anaemia, an autosomal recessive mutation in humans (see Unit 1, Section 1.1), was shown to involve a single amino-acid substitution in the β-polypeptide which altered the solubility of the resultant haemoglobin molecule. This was the very first demonstration that a specific change in a polypeptide was the basis of an observed mutation. Numerous mutations have since been discovered that independently affect either the α- or the β-polypeptide, showing that each polypeptide is specified by a different gene. In general, proteins containing *n different* subunits are specified by *n* genes. We may sum up by stating that a more exact relationship between proteins and genes would be one polypeptide–one gene.

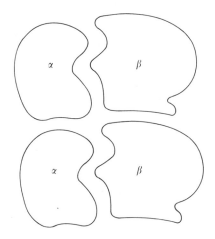

Figure 1 Diagrammatic representation of a haemoglobin molecule with four subunits: two α-polypeptides and two β-polypeptides.

6.1.2 The biochemical basis of complementation

From our knowledge of gene function so far, can we give a more detailed picture of the biochemical basis of complementation?

> **ITQ 2** A certain enzyme is made up of two different polypeptides, A and B. Mutations in either of the two genes specifying the polypeptides can give rise to inactive enzyme molecules. Two individuals, neither of which showed any enzyme activity were mated. One was homozygous for a mutation in the A polypeptide and the other was homozygous for a mutation in the B polypeptide. What kinds of enzyme molecules would be present in the offspring of this mating, and what level of enzyme activity would there be in the offspring compared with normal individuals?

The answer to the previous ITQ indicates that complementation at the biochemical level takes place between mutants defective in different polypeptides (and, perforce, between mutants defective in different proteins). Complementation can often be demonstrated *in vitro* by the formation of normal enzymes in a mixture of extracts from the complementing mutant cells or tissues. Thus, the number of complementing groups (or genes identified) for each enzyme or protein will be the same as the number of different polypeptides that go to make up the functional protein molecule.

Does this mean that whenever complementation occurs, we are to expect the existence of different polypeptides? The answer is that there are exceptions. Complementation can sometimes occur even if the mutations affect the same polypeptide. This is called *intragenic complementation* (to distinguish it from the usual intergenic complementation); it occurs in proteins made up of at least two identical subunits. If the two different mutations involve amino-acid substitutions in different parts of the polypeptide chain, there will be present in the heterozygote two kinds of mutant polypeptides producing hybrid protein molecules that may to some extent function normally and give the appearance of complementation. Such intragenic complementation is often less efficient than complementation between genes, and careful biochemical quantification of enzyme activity (where the protein involved is an enzyme) may distinguish between the two alternatives.

intragenic complementation

6.1.3 Summary of Section 6.1

1 The work of Beadle and Tatum suggested that single metabolic blocks were the result of mutations in single genes. This gave rise to the one enzyme–one gene hypothesis.

2   Biochemical analyses of auxotrophic mutants contributed to working out biochemical pathways involved in the synthesis of essential metabolites.

3   The first direct demonstration that gene mutations altered proteins came from the identification of amino-acid substitution in one of the two kinds of polypeptide chain in the molecule of sickle-cell haemoglobin. This observation necessitated the modification of the one enzyme–one gene hypothesis to one polypeptide–one gene.

4   Many enzyme molecules are made up of aggregates of polypeptide chains. The polypeptide chains are either of different kinds, specified by different genes, or of the same kind, specified by a single gene.

5   At the biochemical level, complementation usually occurs between genes specifying different polypeptides.

6   Rare instances of intragenic complementation, or complementation within a gene, can occur in proteins consisting of identical polypeptide subunits. These arise when mutations lead to amino-acid substitutions in different parts of the polypeptide chain so that two different mutant chains in the heterozygote can associate to give a molecule that is to some extent functional.

7   Biochemical complementation can be demonstrated *in vitro* by the detection of normal enzyme activity in a mixture of mutant polypeptides.

Now try SAQ 1 on p. 286h.

6.2   The structure of genes

In the previous Section, we saw that the gene (as a unit of function) is that which specifies a polypeptide. We also saw that within this gene, a number of mutations can arise which lead to changes in the polypeptide. Until the 1940s, it had been assumed that recombination took place only between genes (i.e. different units of function). In other words, it was assumed that there would be no recombination between mutants that did not functionally complement. Then it appeared that recombination at very low frequencies did occur between non-complementing mutants in both *D. melanogaster* and micro-organisms. In this Section, we shall see how the early molecular geneticists worked out the interrelationships between the units of complementation, mutation and recombination.

Molecular genetics, as we know it today, really began in the early 1950s. First Hershey and Chase provided the definitive proof that DNA *was* the genetic material (see (*HIST*, Section H.5.5), and then in 1953, Watson and Crick worked out the detailed structure of DNA (*HIST*, Section H.5.6). Suddenly it became possible to relate abstract concepts derived from genetic experiments to the molecular properties of the genetic material. In Seymour Benzer's words, as he described his own very important experiments. 'We would like to relate the genetic map, an abstract construction representing the results of recombination experiments, to a material structure . . . DNA' (Benzer, 1957).

6.2.1   Fine-structure mapping and the dissection of genes

The idea underlying fine-structure mapping is that the gene, defined as a unit of complementation, is divisible into smaller units of mutation which can be mapped by recombination. First we shall establish that recombination does occur between mutants that do not complement each other. Then we shall describe the strategy of fine-structure mapping and the way in which the experimental results clarify our ideas concerning the molecular aspects of the gene. As our example we shall use Benzer's elegant work in the 1950s with the phage T4. The starting point of this research was the chance observation that among the *r* (rapid-lysis) mutants was a sub-class that failed to form plaques on *E. coli* K although they formed plaques on *E. coli* B.

QUESTION What are *r* mutants?

ANSWER They are mutants of the bacteriophage T4 that are able to lyse *E. coli* cells much more rapidly than the wild-type (r^+) phage. Thus, they produce large clear plaques in contrast to the small fuzzy plaques produced by the wild-type phage.

Benzer called *r* mutants that were unable to form plaques on *E. coli* K *r*II mutants, and surmised that they must lack some functional protein or proteins specified by some part of the T4 genome. (It may help you to revise details of the lytic cycle of bacteriophage which was discussed in Unit 4, Section 4.7 and is illustrated in *Life Cycles**.) Over the years, Benzer isolated some 3 000 independent spontaneous *r*II mutants. He tested for complementation between pairs of *r*II mutants using the following procedure. About 10^7 particles of a particular mutant phage were mixed with approximately 10^8 cells of *E. coli* K and plated on agar in a petri dish. When the surface hardened, it was overlaid with drops of suspensions containing other mutant phages. In the area occupied by each drop, the two different mutants could infect the same bacterial cell, creating a situation similar to diploidy, in which complementation could occur if the two mutant haploid genomes had different functional deficiencies. The presence of complementation would be indicated by lysis in the area of mixed infection, that is, the restoration of the wild-type ability to lyse *E. coli* K.

QUESTION Match the following observations about the outcomes of such mixed infection with the appropriate conclusions.

Observations
1   Lysis of the K bacteria
2   No lysis of the K bacteria

Conclusions
A   The two mutants fail to complement and therefore possess the same functional deficiencies.
B   The two mutants complement and therefore possess different functional deficiencies.

ANSWER Conclusion A arises from observation 2 and conclusion B from observation 1.

By applying this procedure, it was shown that each *r*II mutant fell into one of two complementing groups, which Benzer designated *r*IIA and *r*IIB. All mutants belonging to one group complemented any member of the other group, but no complementation took place between members of the same group. This delineated two genes, *A* and *B*, which were later shown to be adjacent on the phage chromosome.

Now, Benzer's aim was to discover how finely he could dissect the *r*II region. In other words he wanted to look for the *lowest recombination frequency* between his *r*II mutants and then to translate this into molecular terms. Before we discuss this work, however, let us look at some details of a phage cross.

QUESTION How would Benzer carry out a phage cross?

ANSWER To carry out a phage cross, the two mutants must together infect a host that supports the growth of both mutants. *E. coli* B is such a host.

Let us see what would happen in an *E. coli* B cell infected by both mutants, rII$_1$ and rII$_2$. The two mutant genomes are replicated together in the same cell. Provided that the sites of mutation are not identical, recombination will take place between them to produce the wild-type and the reciprocal double-mutant genomes (see Fig. 2).

* The Open University (1976) S299 LC *Life Cycles*, The Open University Press. This folder, containing details of organisms mentioned in the Course, is part of the supplementary material for the Course.

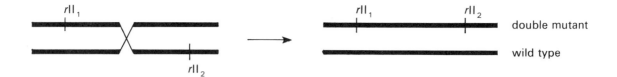

Figure 2 Recombination between two *r*II mutants to give a wild-type and a double mutant.

QUESTION How could the wild-type recombinants be scored?

ANSWER After the phage cross was completed in *E. coli* B, the progeny phage (containing both parental types and recombinants) were plated on to separate bacterial lawns of *E. coli* B and *E. coli* K. The number of plaques formed on *E. coli* K was the number of wild-type recombinants whereas the number of plaques on *E. coli* B was the total number of progeny.

In order to calculate the recombination frequency between the two mutations, $r\text{II}_1$ and $r\text{II}_2$, it is necessary to multiply the number of wild-type recombinants by two. (This is because we must always count the reciprocal recombinant class—in this case the double-mutant recombinants (see Fig. 2) that do not form plaques on K.) The recombination frequency is then

$$\frac{2 \times \text{number of plaques on } E.\ coli\ \text{K}}{\text{number of plaques on } E.\ coli\ \text{B}} \times 100 \text{ per cent}$$

Using the technique described above to select for wild-type recombinants it was theoretically possible to detect recombination frequencies as low as 0.000 1 per cent. Nevertheless, mapping thousands of *r*II mutants using crosses between pairs remained a formidable if not impossible task.

QUESTION How many separate crosses between pairs would be necessary to map 10, 100 and 1 000 different mutants respectively?

ANSWER A general formula can be derived to calculate the number of crosses required. If there are n mutants, each would have to be crossed with every other mutant (i.e. $n - 1$), but each such combination has to be made only once, so the total number of crosses is $n(n - 1)/2$. When n is large, the formula approximates to $n^2/2$. Thus, for 10 mutants $(10 \times 9)/2 = 45$ crosses would be required; for 100, 4 950, and for 1 000 just under 500 000 crosses.

Benzer had nearly 3 000 mutants, so he faced the prospect of carrying out approximately four million crosses to complete his fine-structure map. Fortunately, he was able to use a short cut after he realized the significance of certain observations. He first noticed that although most of his *r*II mutants gave wild-type revertants during growth, a minority did not. He surmised that the reverting class were *point mutants*; **point mutant** that is, they owed their mutant phenotype to the alteration of a single base-pair. As this alteration was presumably a reversible change, the wild-type base-pair would be restored by chance during phage multiplication. The defect in the non-reverting class was clearly different—and the clue to this difference came from another observation. In crosses between pairs, point mutants nearly always gave wild-type recombinants (except where the mutations were at exactly the same point). The non-reverting mutants, on the other hand, failed to give wild-type recombinants when crossed with some point mutants that did give wild-type recombinants in crosses with each other. The non-reverting mutants appeared to have simultaneous alterations of many genetic sites. Benzer concluded that they were, in fact, deletions of parts of the *r*II genome. These *deletion mutants* will not give wild-type recom- **deletion mutant** binants when crossed with phage strains carrying mutations that map within the deleted segment, but will do so with mutants that map outside the deletion (see Fig. 3). The extent of the different deletions varied, and the extent of overlapping between any two deletions could be determined from a cross between them (Fig. 4).

By crossing a number of deletion mutants, a *deletion map* could be constructed which **deletion map** subdivided the *r*II region according to the extent of overlapping between different deletions. Let us follow the steps involved in constructing a deletion map by looking at some of Benzer's very extensive data (see Table 2). The Table records only whether or not recombination has occurred to give wild-type phage, but does not indicate the frequency of recombination. Note the similarity to complementation maps.

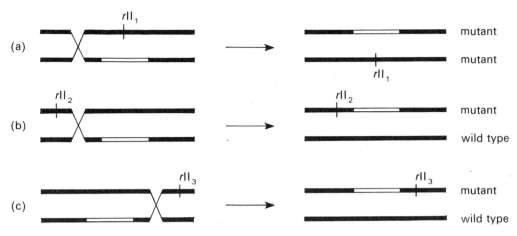

Figure 3 Cross-overs between various point mutants at different positions and the deletion mutant. (a) A point-mutation within the deleted region. (b) and (c) Point mutations outside the deleted region. Note that no cross-over can occur within the deleted region as there is no homologous pairing. The deleted region is represented in white for illustration only. Actually, the chromosome carrying the deletion appears shorter by the amount deleted.

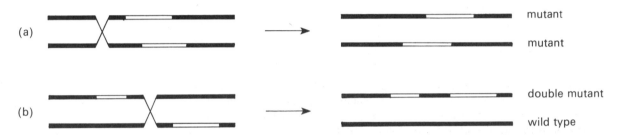

Figure 4 Cross-overs between different deletion mutants. (a) Two mutants with overlapping deletions do not give wild-type recombinants. (b) Two mutants with non-overlapping deletions give wild-type recombinants.

Table 2   The occurrence of wild-type recombinants in crosses between T4 *r*II deletion mutants

| Mutant | Five different deletion mutants (1–5) with their code names | | | | |
	1 (PB28)	2 (1589)	3 (221)	4 (184)	5 (1231)
5	−	−	−	−	−
4	+	+	+	−	
3	−	+	−		
2	−	−			
1	−				

(+ = formation of wild-type recombinants; − = no *r*⁺ recombinants)

Some of these mutants (e.g. 1 and 2) failed to recombine, so their deletions must coincide or at least overlap. Conversely, other pairs of mutants (e.g. 2 and 4) do recombine, so their deletions do not overlap.

QUESTION   Taking each mutant of Table 2 in turn, which mutant(s) does it overlap and which does it not overlap?

ANSWER

| Individual deletion mutant | Mutants | |
	overlapped	not overlapped
1	2, 3, 5	4
2	1, 5	3, 4
3	1, 5	2, 4
4	5	1, 2, 3
5	1, 2, 3, 4	

This pattern of overlapping and non-overlapping can be represented as a map. The convention is to represent each deletion by a horizontal line: the lines that overlap represent an inability to recombine and non-overlapping lines an ability to recombine. For example, consider mutants 1, 4 and 5. Because 5 overlaps both 1 and 4, but 1 and 4 do not overlap, the map for these mutants is drawn as follows:

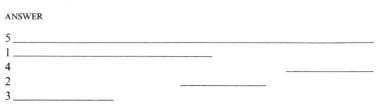

Note that a gap has been left between deletions of mutants 1 and 4 although we do not know if such a gap exists; the reason will become obvious later.

An important point about such a map is that it reveals that the deletions in some mutants are longer than those of others. Here the deletion of mutant 5 is longer than that of either 1 or 4.

QUESTION Can you extend this map, using the same principles, to include mutants 2 and 3?

ANSWER

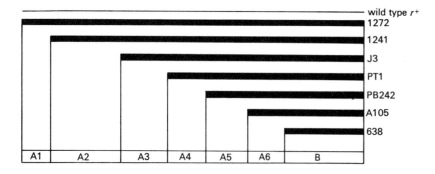

Mutants 2 and 3 overlap mutants 1 and 5, but do not overlap each other or mutant 4. Mutants 2 and 3 have shorter deletions than mutant 1, but again, we do not know if there is a gap between 2 and 3, or for that matter, between 2 and 4. However, we do have the beginning of a deletion map of these 5 mutants*.

Deletion maps can be refined, and ambiguities like those we have already encountered in this exercise can be ironed out through studies using further deletion mutants. For example, the discovery of a deletion mutant that failed to recombine with deletions 2, 4 and 5 but recombined with 1 and 3 would indicate clearly that there was, in fact, a gap between deletions 1 and 4.

In this way Benzer was able to subdivide the rII region into sections, each section defined by the length of the map covered by one deletion but not by another. This gave a rudimentary deletion map of part of the rII region (see Fig. 5).

Figure 5 A rudimentary deletion map that subdivides the rII region of phage T4 into seven parts: A1 to A6 in gene *A*, and B in gene *B*. Note that in the deletion map the deleted segments are represented in thick black lines.

Now Benzer was in the position to carry out deletion mapping in order to reduce the number of crosses between pairs he needed to make to map the rII region. Figure 6 shows the map of the rII deletions that Benzer finally used to sort out his mutants.

In order to map a mutant, he first crossed it to each of the seven major deletion mutants (illustrated in Fig. 5 and shown at the top of Fig. 6). Having located the mutation within one of the major sections, A1, A2, A3, A4, A5, A6 or B, he then crossed the mutant to a secondary set of deletions. Thus, if the mutation were initially located in section A1, it could subsequently be crossed with deletion mutants 1364 and EM66 to see if it was present in the subdivisions, A1a, A1b1 or A1b2.

* The mirror image of this map (i.e. with deletion 1 to the right of 4) is also consistent with the data of Table 2; but this does not matter at this stage in our exercise.

ITQ 3 A new *r*II mutant was located within the A1 region. It failed to give wild-type recombinants with both 1364 and EM66. In what subdivision of the A1 section is the mutation located (see Fig. 6)?

Figure 6 A detailed deletion map of the *r*II region of phage T4 showing 47 subdivisions. Note that some ends of the deletions are not used to define segments; these are shown as fluted.

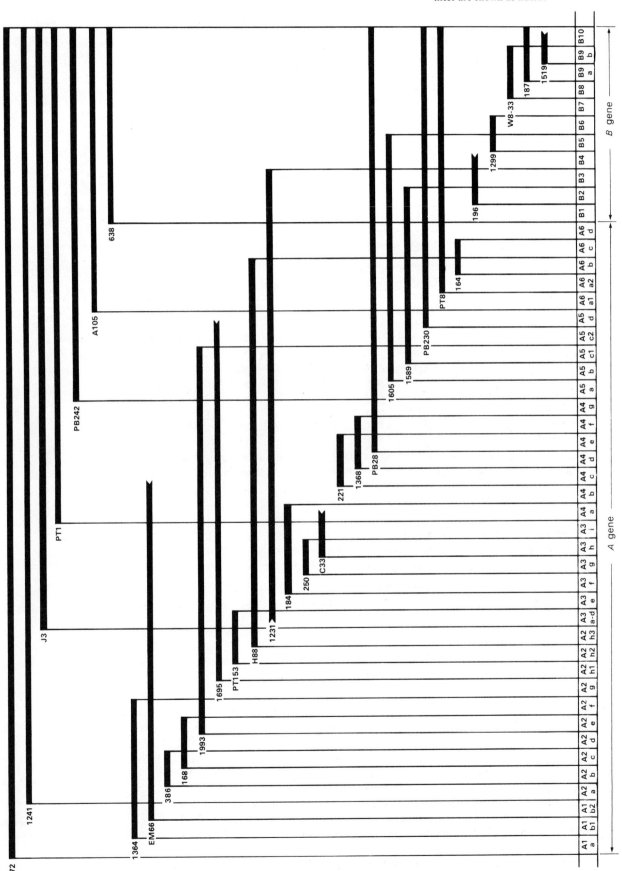

Thus, with only two sets of experiments, any point mutation could be placed in one of the 47 subdivisions of the *r*II genome. It would then be a simple matter to make a few crosses between point mutations within the same region in order to establish the recombination frequencies between each point mutation. Compare this strategy with the theoretical necessity of making millions of crosses!

Over 2 000 different *r*II mutants were mapped by this combination of deletion mapping for approximate location, and subsequently each was crossed with selected point mutants for more accurate location within the subdivision of the *r*II genome. Now test your skill in deletion mapping with the following ITQ.

ITQ 4 The results of crosses between a new *r*II mutant and two sets of deletion mutants included in Figure 6 were:

	Mutant code numbers						
Deletion set 1	1272	1241	J3	PT1	PB242	A105	638
	−	−	−	−	−	+	+
Deletion set 2	PB28	1605	1589	PB230			
	−	−	+	+			

($+$ = formation of wild-type recombinants; $-$ = no r^+ recombinants)

In which segment of the map is the new mutant located? (Work through the results of deletion set 1 first and then through those of set 2.)

6.2.2 The fine-structure map of the *r*II region

Benzer's results showed that the *r*II region consists of a total of ten map units covering two adjoining genes, *r*II*A* and *r*II*B*. The lowest recombination frequency obtained in the crosses was 0.01 per cent. How do these results translate into molecular terms? It is known that the entire T4 phage genome is about 1 500 map units in length, and that it contains about 2×10^5 nucleotide pairs. Thus, the size of the *r*II region can be calculated as $10/1\,500 \times 2 \times 10^5$ or about 1 300 nucleotide pairs. So the lowest recombination frequency of 0.01 per cent represents a distance of less than two base pairs. This is consistent with the idea that recombination can take place between adjacent nucleotide pairs.

Figure 7 (opposite) shows the detailed map obtained by Benzer for his spontaneous mutants. The map is linear right down to the smallest level (although it has been bent to fit the page!). Each unit square represents one mutant located at that particular site. As can be seen, mutations do not occur with equal frequencies throughout the *r*II region; some points have many more mutations than others. These are called '*hot-spots*'.

hot-spots

Benzer's work showed how the units of complementation, mutation and recombination relate to each other. Thus, the complementing unit appears to be complex; numerous mutations occur within it, and these mutations can be recombined. In terms of DNA, the complementing unit consists of hundreds of nucleotide pairs, whereas the smallest mutating unit is probably a single nucleotide pair, and therefore the smallest recombination unit must be the distance separating two adjacent nucleotide pairs.

6.2.3 Summary of Section 6.2

1 A gene defined by complementation (according to function) is shown to be different from that defined by mapping (recombination).

2 Fine-structure genetic maps of closely linked mutants that do not complement each other can be constructed by fine-structure mapping. We restrict ourselves to considering *r*II mutants of phage T4.

3 In mapping the *r*II region, Benzer employed the following method: (a) complementation tests to separate mutants into two functional groups A and B; (b) the

262

construction of deletion maps to subdivide the *r*II region; (c) deletion mapping to locate point mutations within each subdivision; and (d) two-point crosses between point mutations within a subdivision in order to determine the recombination frequencies between each adjacent mutant. This method greatly simplified the task of mapping thousands of mutants by conventional techniques which would have entailed millions of two-point crosses.

4　The map Benzer constructed covered 10 map units corresponding to some 1 300 nucleotide pairs. Recombination was estimated to take place between adjacent nucleotide pairs.

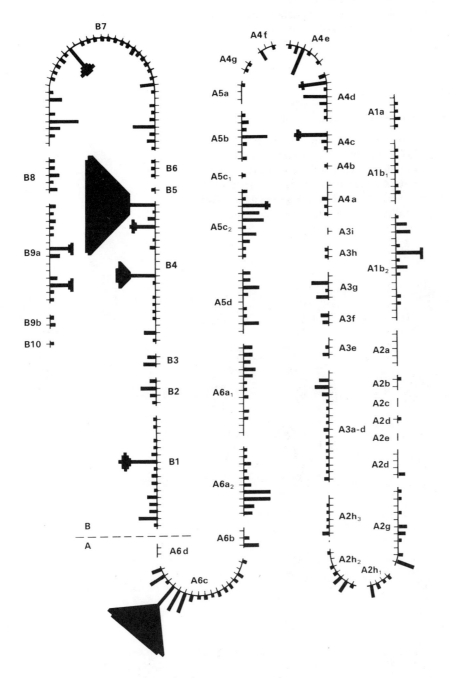

■ area represents one mutant

Figure 7　A map of spontaneous *r*II mutants of phage T4.

5　The map of *r*II spontaneous mutants (Fig. 7) revealed that mutations are not randomly distributed throughout the *r*II region. Some points (hot-spots) are more susceptible to mutation than others.

Now try SAQs 2, 3 and 4 on p. 286h.

6.3 Mutagenesis and the molecular basis of mutations

Mutagenesis is the induction of mutations by chemical or physical means. Remember that mutations are observed as phenotypic differences from the wild type; it is necessary, therefore, to discover the nature of the alterations in the genetic material that give rise to these differences. In the previous Section, you saw that Benzer's experiments in fine-structure gene mapping strongly suggested that mutations could involve single nucleotide changes in DNA. If this were the case, then chemicals that alter bases in the DNA would be expected to cause mutations. Also, as different chemicals exhibit different specificities in their action, a corresponding specificity of mutation should result from the application of each chemical. A study of mutagenesis should therefore throw light on the molecular basis of mutations. Such were the considerations which led to the study of chemical mutagenesis.

6.3.1 Chemical mutagenesis

If you do not understand the chemical terms in this Section, refer to S100 (S100²). (*Note*   You are not required to memorize any chemical formulae.)

At the time of the discovery of the structure of DNA in 1953, Watson and Crick suggested that spontaneous mutations could result from illegitimate pairing of bases during DNA replication. To refresh your memory of base pairing, there are four bases in DNA: two purines, adenine (A), and guanine (G); and two pyrimidines, thymine (T) and cytosine (C). Pairing occurs by specific hydrogen bonding between a purine and a pyrimidine. Thus, adenine pairs with thymine, and guanine with cytosine (see Fig. 8(a)). Both thymine and adenine can exist in freely interconvertible forms known as tautomers.

(a)

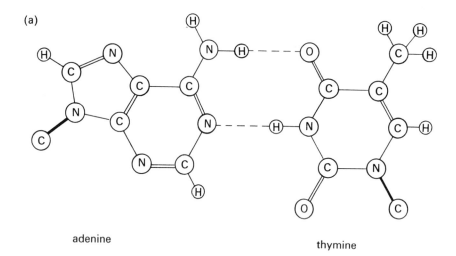

adenine   thymine

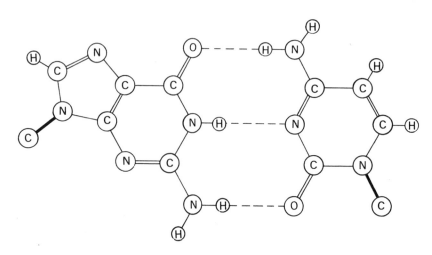

guanine   cytosine

Figure 8(a) Normal base pairing in DNA: adenine with thymine; guanine with cytosine.

264

(b)

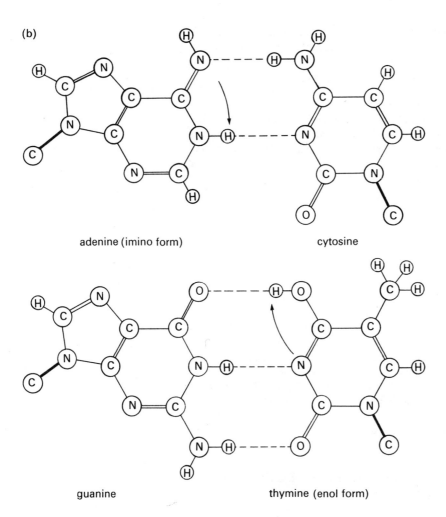

adenine (imino form)   cytosine

guanine   thymine (enol form)

Figure 8(b) Mispairing involving rare base tautomers: the imino form of adenine pairs with cytosine and the enol form of thymine pairs with guanine.

Rare tautomers of these bases can pair illegitimately (see Fig. 8(b)). Thus, the imino form of adenine (with the $=$NH group) pairs with cytosine, and the enol form of thymine (with the $=$C$-$OH group) pairs with guanine. In Figure 9 we see that the result of such *mispairing* in the DNA molecule during the first replication (A erroneously paired with C) is that an original A$-$T pair becomes a G$-$C pair after the next replication.

mispairing

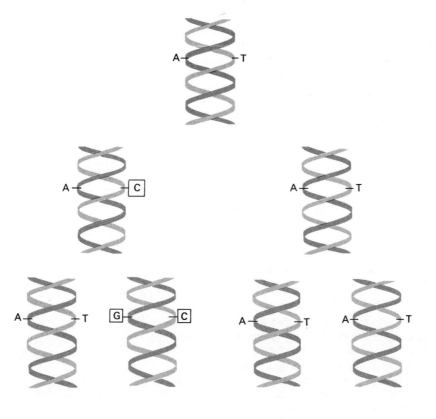

Figure 9 Mispairing in DNA replication gives rise to base-pair substitution in the next replication. The substituted bases are enclosed in boxes.

Certain chemical base analogues were tested for their ability to induce mutations. It was reasoned that mispairing would be more frequent if the analogues were incorporated into DNA in place of the normal bases because the hydrogen bonding would be less precise. The first analogues tested were 5-bromouracil, a pyrimidine analogue capable of replacing thymine in DNA, and 2-aminopurine, a purine analogue capable of replacing adenine. The tests for mutagenicity were carried out in the rII system of phage T4. Each chemical was tested for its ability to induce rII mutants in r^+ strains, and the mutations induced were mapped by fine-structure mapping. It was found that each base analogue gave rise to a special distribution of hot-spots and other sites of mutation. This suggested that the base analogues were being incorporated into newly synthesized DNA strands; this caused mispairing which in turn gave rise to mutations by *base substitution*.

mutagens

base substitution

ITQ 5 Reconstruct the sequence of events leading to base substitutions starting from the incorporation of each base analogue.

Base substitutions caused by base analogues are called *transitions*; they involve a replacement of purine for purine (A ↔ G), or pyrimidine for pyrimidine (T ↔ C). The replacement of a pyrimidine by a purine, or vice versa (A ↔ T, G ↔ C, A ↔ C and G ↔ T) is known as a *transversion*. In practice, both 5-bromouracil and 2-aminopurine cause the transition of an A—T pair to a G—C pair as well as the reverse because of the ease with which they can undergo illegitimate pairing. Thus, 5-bromouracil can become incorporated into DNA in place of cytosine and lead ultimately to the replacement of a G—C pair by an A—T pair. Proof that the mutations induced were single base-pair substitutions came from the observation of frequent spontaneous reversions to the wild type. The frequency of reversions would be expected to be increased by base analogues if the original substitutions were transitions.

transition

transversion

ITQ 6 What effect do base analogues have on the reversion rates of mutants first induced by base analogues?

Table 3 summarizes the actions of some chemical mutagens.

Table 3 A summary of chemical mutagens and their effects

Mutagen	Mode of action	Effect on nucleic-acid base pairs
5-bromouracil 2-aminopurine	analogues substituting for normal bases	two-way transitions (A—T ⇌ G—C)
nitrous acid	deamination of bases	two-way transitions (A—T ⇌ G—C)
hydroxylamine		one-way transitions (G—C → A—T)
methyl- and ethyl-methanesulphonate	alkylation of bases	two-way transitions (A—T ⇌ G—C)
nitrogen mustards		and transversions (A—T ⇌ T—A or C—G)
nitrosoguanidine		
acridines (e.g. proflavin)	distortion of double helix	insertion or deletion

Two of these classes of chemicals are known to react with bases. These are:

1 Deaminating agents that remove amino groups from the bases, for example, nitrous acid and hydroxylamine. The removal of amino groups from normal bases encourages illegitimate pairing and induces transitions.

2 Alkylating agents which add alkyl groups to bases, making them unstable and liable to chemical breakdown. These include diethylether, methyl- and ethylmethane-sulphonates, nitrogen mustards and nitrosoguanidine. The alkylating agents vary in their action; some like ethylmethanesulphonate and nitrosoguanidine induce only transitions, others like methylmethanesulphonate can induce both transitions and

transversions. Transversions will occur when the base is destroyed *in situ*, so that any one of the other three bases can replace it.

The mutants induced by these classes of chemical agents can revert spontaneously, and the addition of base analogues increases the reversion rate.

Yet another class of mutagens are the acridine dyes such as proflavin and acridine orange. These dyes probably intercalate between base pairs in the double helix, thus distorting the structure of the DNA and ultimately causing a deletion or an insertion of bases. Mutants induced by acridine dyes revert much less frequently and the rate of reversion is not increased by the addition of base analogues.

6.3.2 Radiation mutagenesis

Irradiation with X-rays has been used for at least 50 years for the induction of mutation (Unit 2, Section 2.11.4). The mutagenic action of X-rays is not understood; they induce frequent chromosome breaks (see Unit 5) resulting in deletions, though point mutations (or base substitutions) could also occur.

Other radiations that cause mutations are ultraviolet light, beta and gamma rays. Ultraviolet light can induce single base substitutions whereas beta and gamma rays are less specific, often giving rise to a wide variety of changes similar to those induced by X-rays.

6.3.3 Spontaneous mutations

The molecular basis of spontaneous mutations is less well known. You have met some examples of spontaneous mutations in this Unit. For example, Benzer's T4 *r*II deletion mutants and point mutants were all spontaneous (Section 6.2.1). The largest deletion involved some 1300 base pairs.

When spontaneous point mutations were mapped by fine-structure mapping, the distribution differed from those induced by mutagens, and the pattern of occurrence of hot-spots was unique. There is some evidence that spontaneous mutations are often deletions ranging in size from one to several hundred bases. Base substitutions also occur, but most of them are transversions. It seems unlikely therefore, that mispairing during DNA replication contributes significantly to spontaneous mutations.

The DNA of a cell is subject to a variety of natural mutagenic agents at all times, notably the ultraviolet component of sunlight, for example. Yet mutations accumulate relatively slowly. The reason for this is the presence of a DNA repair system in each cell. This consists of a group of enzymes that work in sequence to remove mismatched bases, to insert new ones and join up loose ends. If the repair system completes its tasks before DNA replication (when base changes become established), there is, in effect, no mutation.

6.3.4 The molecular basis of mutations

The evidence presented in this Section indicates that mutations involve one of the following changes to DNA:

(a) single base-pair substitutions;

(b) the insertion of base pairs;

(c) the deletion of base pairs.

These lead to a change in the base sequence of DNA which in turn affects the function of the gene. In Section 6.1 we saw that genes specify polypeptides and that mutants can be identified by an alteration in the polypeptide. In this Section, we have seen that mutations involve a change in the base sequence of DNA. In the next Section, we shall examine the way in which genes specify polypeptides; that is, we shall investigate the coding relation between DNA and polypeptides.

6.3.5 Summary of Section 6.3

1 Mutations can arise due to mispairing of bases during DNA replication. This results in single base substitutions.

2 Single base substitutions are of two kinds, transitions (substitutions of purine for purine, or pyrimidine for pyrimidine) or transversions (substitutions of purine for pyrimidine or vice versa). Analogues of normal bases, deaminating and alkylating agents, can also cause transitions; some alkylating agents cause transversions.

3 Acridine dyes induce mutations by the insertion or deletion of base pairs.

4 Radiation, such as X-rays, beta and gamma rays, causes mutations by deletions of base pairs or single base-pair substitutions. Ultraviolet light mutagenesis involves single base-pair substitutions.

5 Spontaneous mutations include both single base-pair substitutions and deletions.

6 The study of mutagenesis shows that the molecular basis of mutations is a change in the base sequence of DNA.

Now try SAQ 5 on p. 286h.

6.4 The genetic code

So far, we have seen that

(a) the gene as a unit of function specifies a polypeptide,

(b) the unit of function corresponds to a stretch of the DNA molecule, and

(c) changes in the base sequence of DNA can lead to mutations, some of which have been identified as changes in the amino-acid sequence of the polypeptide (see Section 6.1.1).

These facts tell us that there is some sort of information contained in the base sequence of DNA that is somehow transferred to the polypeptide, determining its sequence of amino acids. In other words, there must be a coding relationship between the stretch of the DNA molecule and the polypeptide it specifies.

From your reading of S100 (S100^3) you already know that the information contained in the DNA is transcribed on to a messenger RNA molecule; this then passes to the cytoplasm where it is translated into a sequence of amino acids joined end-to-end to form a polypeptide. The details of this process are revised in Appendix 1 (p. 286f). *You are strongly advised to familiarize yourself with them before proceeding with the remainder of this Unit.*

Let us look at the problem of coding. There are 20 different amino acids and these have to be specified by combinations of the 4 bases.

QUESTION Consider the possibility that: (a) one; (b) two; (c) three; and (d) four bases are required to code for each amino acid. What is the total number of amino acids that could be coded for in each case?

For (b), (c) and (d) assume that:

(i) any combination of bases is acceptable (including the situation in which all the bases are the same);

(ii) reverse sequences of bases have different meanings (e.g. BA is different from AB).

ANSWER (a) 4; (b) $4^2 = 16$; (c) $4^3 = 64$; (d) $4^4 = 256$.

QUESTION Which of these answers is consistent with 4 bases coding for 20 amino acids?

ANSWER (a) and (b) are not consistent because in neither is the maximum coding potential more than 16, and there are 20 amino acids. (c) and (d) are certainly adequate, more than adequate, in fact! Triplets of bases could code for more than 3 times the actual number of amino acids naturally occurring in proteins, quadruplets for more than 10 times that number. In both (c) and (d) the code could be *degenerate*, that is, single amino acids could be coded for by more than one triplet or quadruplet. On grounds of cellular economy, the extent of degeneracy in a quadruplet code makes such a code appear unlikely, even before any experiments are carried out.

degeneracy

So it seems we have a triplet code, in which a sequence of three bases codes for one amino acid. Such a code could be overlapping or non-overlapping. In other words, amino acids could be read off one after another from sequences of three bases (along a stretch of the DNA) which either overlapped or not (Fig. 10). In an overlapping code, each base on the DNA could participate in the coding of two or

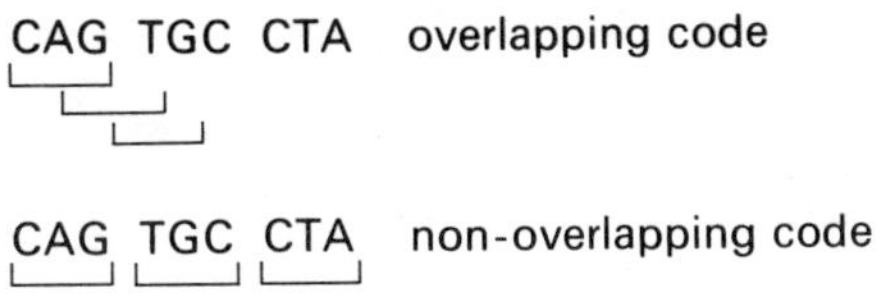

CAG TGC CTA   overlapping code

CAG TGC CTA   non-overlapping code

Figure 10

three amino acids, but this would impose severe restrictions on the possible sequences of amino acids that could occur in proteins. Thus, for instance, an amino acid whose code was CAG could not be followed by one with the code in which the first base was T or C because, by definition, the first base of the next amino acid would have to be A or G, depending on the extent of overlapping. It was Brenner who first proposed that the code was non-overlapping, so that each base participated in the coding of only one amino acid. The triplet of bases specifying one amino acid was called a codon.

In the radio programme, The *r*II system: solving the code, Francis Crick talks about the way in which he proved the existence of the triplet code.

6.4.1   The triplet nature of the code

When Benzer was working on the *r*II system of T4 phage, he noticed that the point mutants quite frequently reverted to the wild type. You may have assumed that this was because of a reversal of the original mutation which resulted in the wild-type sequence of bases again. Such true reversions are not the only way in which *r*II point mutants can regain the wild-type phenotype. It was found that, in some instances, the revertants actually carried a second mutation that had occurred close to the original mutation. The two mutations together were responsible for producing the wild-type phenotype—that is, the second mutation suppressed the first mutation, and was therefore, not surprisingly, called a *suppressor mutation*. Revertants which carried such suppressor mutations were *pseudo wild-type*. Francis Crick and Sydney Brenner were able to exploit suppressor mutations to prove for the first time that the genetic code was really composed of triplets, 8 years after the idea had first been proposed. They were also able to demonstrate that the DNA base sequence was read from a fixed point, triplet by triplet, in a non-overlapping manner. This meant that each genetic message had a fixed *reading frame* which could be altered by, say, adding or deleting a single base. Whenever this occurred, the polypeptide specified by the message would contain incorrect amino acids starting from the point at which the base was added or deleted. Their work also strongly suggested that the code was degenerate.

suppression
pseudo wild-type

reading frame

Crick and Brenner began by demonstrating that most of the apparent wild-type revertants of a T4 *r*II mutant called FCO were not in fact true reversions of the original mutation, but were double mutants because they carried a suppressor mutation very close to the original *r*II mutation. The combination of these two mutations permitted growth on *E. coli* K, giving the wild-type phenotype. Crick and Brenner proved the existence of the suppressor mutation by crossing the pseudo wild-type with the authentic T4 *r*⁺ wild-type phage. Figure 11 shows how this was done. FC7 is the name of the suppressor mutation.

If the pseudo wild-type had been a genuine revertant from the original mutation then only *r*⁺ progeny would have been produced in this cross.

Crick and Brenner then showed that the *r*II mutant FC7, which had been isolated when they crossed the pseudo wild-type with the *r*⁺ phages, could also revert to give

Figure 11 A cross-over between a pseudo wild-type and a wild type T4 phage gives rise to two mutant recombinants.

a wild-type phenotype. However, when these revertants were crossed with the authentic T4 r^+ stocks many of them turned out also to be pseudo wild-type, produced by the combination of FC7 with yet another mutation. Thus this latest mutation was a suppressor of a suppressor mutation. It was also possible to isolate suppressors of suppressors of suppressors!

The important feature in these experiments, however, was that these mutants had been induced by acridine dyes.

QUESTION What is the normal mode of action of acridine dyes on DNA?

ANSWER They cause the *insertion* or *deletion* of bases in the DNA nucleotide sequence.

base insertion
base deletion

Crick and Brenner made the assumption that the FCO mutation was actually an insertion of just one base in the DNA sequence of the *r*II region. Thus, if the DNA base sequence was being read triplet by triplet (from left to right), starting from a fixed point, the insertion of a base would cause all the triplets to the right of the FCO mutation to be read out of frame—it would be a *frameshift mutation*. This would presumably result in a non-functional polypeptide (Fig. 12).

frameshift mutant

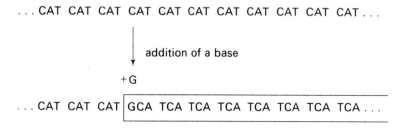

Figure 12 The insertion of one base shifts the reading frame.

They next postulated that if the FCO mutation was indeed an insertion then the FC7 mutation must have been a deletion. See how the combination of the two of them would affect the reading frames—still assuming that the triplets are read left to right (Fig. 13).

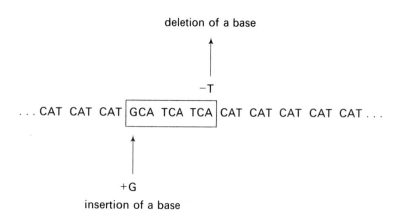

Figure 13 The insertion of one base followed by the deletion of another base restores the reading frame.

The correct reading of the triplets would be interrupted after the FCO mutation, but it would be restored after the FC7 mutation. Thus, providing that the resultant polypeptide is not totally inactivated by this small section of altered amino acids, there is no reason why it should not be functional. The FC7 mutation has been found to map very close to the FCO mutation—presumably suppressors are always fairly close to the mutation they suppress as it is unlikely that a polypeptide with a large number of altered amino acids could remain functional.

Note that it was not *known* that the FCO mutation was the insertion of a base; it was just assumed. However, it is essential to the argument that if it *was* an insertion then the FC7 mutation would have to be a deletion. If FC7 had been another insertion it would not have 'cancelled out' the FCO mutation—see Figure 14 (opposite).

Insertions are generally represented by a plus sign (+) and deletions by a minus sign (−), and we can now state that whatever the nature of the original mutation its suppressor is always of the opposite sign.

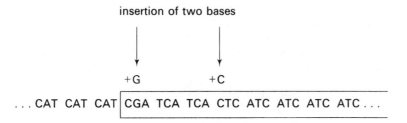

Figure 14 The insertion of two bases fails to restore the reading frame.

Combining two mutations of the same sign (i.e. two insertions, or two deletions) never resulted in the reversion to the wild-type phenotype.

Up to this point we have not yet presented any proof of the existence of a triplet code, we have shown only that it is most likely to read in one direction starting from a fixed point, and that it takes two or more bases to code for each amino acid. It was only when mutants containing three insertion mutations or three deletion mutations were made that the triplet nature of the code could be demonstrated. These triple mutants were found to have wild-type phenotypes (Fig. 15).

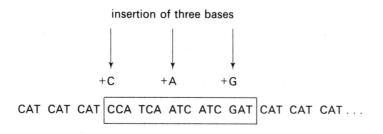

Figure 15 The insertion of three bases restores the reading frame.

This could have occurred only if the code was read in triplets, as only then would the reading frame be re-established after three alterations of the same sign.

The experiments also revealed that it was unlikely that only 20 of the possible 64 codons were used to specify amino acids, for this would have meant that there would be 44 useless codons, none of which corresponded to an amino acid. One of these useless codons would almost certainly be produced between an insertion and its suppressing deletion, and could have caused the termination of the polypeptide before the reading frame was restored.

6.4.2 Codon assignment

While the triplet nature of the code was being demonstrated, scientists had already begun the task of *codon assignment*. In the early days (1961–63) this was done by introducing a synthetic RNA of known nucleotide sequence into a cell-free protein-synthesizing system in a test tube.

codon assignment

> QUESTION What would be the essential constituents of such a system?
>
> ANSWER Ribosomes, tRNAs, aminoacyl-tRNA synthetases (the enzymes that attach the amino acid to the tRNA), ATP and amino acids.

This *in vitro* system treated the synthetic RNA as though it were mRNA and produced a polypeptide by translating its base sequence. The polypeptide was then harvested and analysed to discover the nature and order of the amino acids incorporated into it. Nirenberg and Matthaei were the first to use the technique and they discovered that an RNA containing only the base uracil (poly U) produced a polypeptide composed entirely of the amino acid phenylalanine. They concluded that the triplet code for phenylalanine was UUU. The introduction of poly C and poly A produced polypeptides with only proline and lysine respectively. The synthesis of poly G presented technical problems and therefore it could not be tested in the *in vitro* system.

Nirenberg and Matthaei, and also Ochoa, next produced RNAs containing random sequences of two or more bases. As it was impossible to predict the sequence in which the bases had been linked together, precise codon assignment could not be achieved.

For example, a random sequence of equal numbers of uracil and adenine bases would produce the eight possible triplets—UUU, AAA, UAA, AAU, UUA, AUU, UAU, AUA—in equal frequencies. Thus, polypeptides containing equal amounts of eight amino acids would be produced in the protein-synthesizing system, and although this would give some information about the codons of these amino acids, it would not be possible to assign the particular triplets to the amino acids involved. A way round this difficulty was to mix the bases in unequal amounts; then the various triplets would be produced in unequal quantities, and when the synthetic RNA was introduced into the system, varying amounts of each of the amino acids would be incorporated, depending on how frequently the appropriate triplet was present. In some instances it was thus possible to make definite codon assignments.

Khorana made the next major contribution by synthesizing alternating copolymers such as –UGUGUGUGUG–; this was an outstanding technical achievement. Poly UG was discovered to code for –valine–cysteine–valine–cysteine–.

QUESTION What are the codons suggested for valine and cysteine?

ANSWER GUG or UGU.

Khorana went on to make polyribonucleotides of repeated trinucleotides such as –UUGUUGUUGUUG–, which was found to code for three types of polypeptide— polyleucine, polycysteine, and polyvaline.

ITQ 7 Given the information in Table 4, assign amino acids to each of the codons produced by any combination of U and G.

Table 4 The *in vitro* synthesis of polypeptides using synthetic polyribo-nucleotides (mRNAs)

	Polyribonucleotides	Amino acids of polypeptides
1	random UG(U$_2$G or UG$_2$)	valine, leucine, cysteine, tryptophan, glycine
2	alternating copolymers UGUGUGUG...	valine, cysteine–valine, cysteine
3	repeating trinucleotides UUGUUGUUG...	polyleucine, polycysteine, polyvaline

In 1963 Nirenberg made the next leap forward, which allowed a direct assignment of codons, by discovering that the addition of an RNA of just three nucleotides—a trinucleotide—to the protein-synthesizing system resulted in the binding of a tRNA (with its specific amino acid attached) to the ribosomes. These ribosome–aminoacyl-tRNA complexes could then be easily separated from the reaction mixture by a filtering process.

In practice, a trinucleotide of known base sequence was added to the protein-synthesizing system which contained a mixture of amino acids, only one of which was radioactively labelled. On incubation the aminoacyl-tRNA containing the sequence complementary to the trinucleotide attached itself to the ribosome that had already bound the trinucleotide, treating it as though it were a mRNA. The mixture was passed through a filter, and all the aminoacyl-tRNAs (except those bound to the ribosomes) passed into the filtrate. If the labelled amino acid remained on the filter then its codon was known to be the trinucleotide used in the experiment.

Within a year amino acids had been assigned to most of the 64 possible codons. Eventually, only the codons UAA, UAG and UGA remained unidentified. The final result of all these efforts is given in Table 5. Remember that a codon is a sequence of bases on a mRNA *not* on the DNA. The anticodon is contained in the tRNA that is complementary to each codon.

Three points of particular interest emerge from a study of Table 5.

1 The code is degenerate. Almost all the amino acids have more than one codon. Three have six possible codons, five have four codons, and nine have two alternative codons. Only two amino acids have just one codon—methionine and tryptophan.

2   The code contains three codons for which no amino acids could be found. These were therefore called *nonsense codons*; they are UAA, UAG, and UGA. You will see later that they are in fact punctuation codons giving the 'stop' instruction.

nonsense codon

Table 5   The assignment of codons

First position (read down)	Second position (read across)				Third position (read down)
	U	C	A	G	
U	phe	ser	tyr	cys	U
	phe	ser	tyr	cys	C
	leu	ser	*stop*	*stop*	A
	leu	ser	*stop*	trp	G
C	leu	pro	his	arg	U
	leu	pro	his	arg	C
	leu	pro	gln	arg	A
	leu	pro	gln	arg	G
A	ile	thr	asn	ser	U
	ile	thr	asn	ser	C
	ile	thr	lys	arg	A
	met (*start*)	thr	lys	arg	G
G	val	ala	asp	gly	U
	val	ala	asp	gly	C
	val	ala	glu	gly	A
	val	ala	glu	gly	G

ala—alanine; arg—arginine; asn—asparagine; asp—aspartate; cyc—cysteine; gln—glutamine; glu—glutamate; gly—glycine; his—histidine; ile—*iso*leucine; leu—leucine; lys—lysine; met—methionine; phe—phenylalanine; pro—proline; ser—serine; thr—threonine; trp—tryptophan; try—tyrosine; val—valine.

3   Alternative codons for a particular amino acid normally have identical bases in their first two positions. The obvious exception to this is where there are six possible codons for an amino acid—in this case four of the codons have their first two bases in common.

> **ITQ 8**   An inactive enzyme isolated from a mutant *E. coli* strain was found to have a threonine in place of lysine at one position in its polypeptide chain. Spontaneous revertants of this mutant back to the wild-type were obtained, but when the amino-acid sequences were determined, the revertants were found to have either serine or methionine at the position originally occupied by lysine. What can you say about the mutational changes involved?

6.4.3   The colinearity of the DNA base sequence and the polypeptide chain

When Crick and Brenner's work on the non-overlapping code was discussed we made the assumption that the DNA base sequence was *colinear* with the sequence of amino acids in the polypeptide it coded for—that is, the sequence of triplets coding for amino acids was the same as the sequence of amino acids in the polypeptide. No evidence for the truth of this assumption was produced until 1966 when Yanofsky was able to demonstrate colinearity between his fine-structure map of the *trp* A gene in *E. coli* and tryptophan synthetase A, the polypeptide it coded for.

colinearity

He had isolated a number of *E. coli* mutants, all of which produced a defective tryptophan-synthetase A polypeptide, and had constructed his fine-structure genetic map by making a series of crosses between all the possible pairs.

> QUESTION   From your knowledge of genetic mapping in *E. coli*, which of the three techniques—conjugation, transduction and transformation—could be used to work out the recombination frequencies between the *trp* A mutants?

ANSWER Transduction, because the sites of mutation all affect the same polypeptide and must therefore be very close together. Conjugation would not be suitable for mapping mutations at this sort of distance apart. Transformation was not known in *E. coli* at the time of Yanofsky's experiments. If you are in doubt about the answer to this question, look back to Unit 4 (Section 4.7).

Yanofsky next isolated the defective tryptophan-synthetase A polypeptide from several of the mutants and determined the amino-acid sequence in each case. The fine-structure genetic map was then lined up with the amino-acid map produced for each mutant. It was found that there was an abnormal amino acid at some point in the protein made in each of the mutants and he was able to show that *the order of the mutant sites on the fine-structure genetic map corresponded to the order of the amino acids altered in the various mutant proteins* (see Fig. 16).

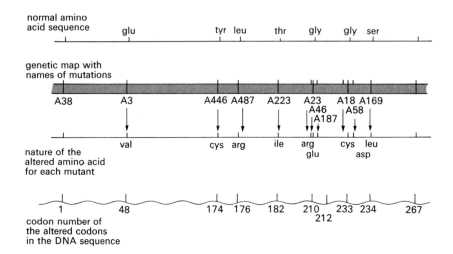

Figure 16 The colinearity of the genetic map and the amino-acid map in *E. coli* tryptophan synthetase A protein.

Find the mutation A46 on the genetic map. The defective tryptophan synthetase A isolated from *E. coli* with the A46 mutation had glutamic acid as its 210th amino acid instead of the glycine found in the wild types. Now locate the A23 mutation. It maps alongside A46. Analysis of the protein from this mutant showed it also had an altered 210th amino acid, but contained arginine in this position rather than either glycine or glutamic acid.

QUESTION What are the codons for glycine (see Table 5)?

ANSWER GGU, GGA, GGC or GGG.

QUESTION What are the codons for glutamic acid?

ANSWER GAA or GAG.

QUESTION What are the codons for arginine?

ANSWER CGU, CGA, CGC, CGG, AGA or AGG.

If we put these three sets of codons together it can be easily seen that a mutation in the first base of the glycine codon could give rise to arginine, whereas a mutation in the second base could give glutamic acid as follows:

$$CG\frac{A}{G} \quad \longleftarrow \quad GG\frac{A}{G} \quad \longrightarrow \quad GA\frac{A}{G}$$

arginine   glycine   glutamic acid

The A46 and A23 mutants are thus caused by mutations in adjacent bases. Yanofsky proved this by producing *trp*⁺ recombinants when the mutants were crossed in a transduction experiment. This gives you an idea of the resolving power of such an experiment, which is able to show recombination even between adjacent bases!

274

Not only does Yanofsky's work demonstrate the colinearity of the triplet sequence and the amino-acid sequence specified by it, but it also shows that a codon that has been altered by mutation in the middle of a polypeptide often results in the insertion of another amino acid. A mutation of this type is called a *missense mutation*, and results in a polypeptide with a single amino-acid substitution.

missense mutation

6.4.4 The direction in which the code is read

Although we have proved that the code is read from a fixed starting point we have *assumed* so far that it is read in only one direction. It was Streisinger and his colleagues who in 1966 actually produced the proof for this assumption. He was working with the frameshift mutants used earlier by Crick in demonstrating the triplet nature of the code.

QUESTION What is a frameshift mutant?

ANSWER A mutant in which insertion or deletion of nucleotide base(s) has caused the triplet to be mistranslated from the point of the mutation onwards.

Streisinger was also using the T4 phage, but he was interested in frameshift mutants that produced defective lysozyme (an enzyme that breaks down the bacterial cell wall), and were thus incapable of lysing the bacteria they infected. Starting from one such frameshift mutant, he isolated a revertant which had regained the ability to make active lysozyme. He then extracted and purified the lysozyme produced by this revertant, and compared its amino-acid sequence with that in lysozyme produced by the wild-type phage.

The two sequences were:

wild type:	–thr–lys–	ser–pro–ser–leu–asn	–ala–ala–lys–
double mutant:	–thr–lys–	val–his–his–leu–met	–ala–ala–lys–

This showed that five amino acids in the two sequences were different as the result of two mutations, presumably an insertion and a deletion of a single base. He worked out all the possible codons for the amino acids affected, and discovered that *only one* sequence of bases in the wild type could have produced the change in amino-acid sequence in the double mutant. This proved that the sequence of triplets was read in one direction only, otherwise it would have been impossible to pinpoint a unique sequence of bases that accounted for the results of the double mutation (Fig. 17).

wild type	thr	lys	ser	pro	ser	leu	asn	ala	ala	lys
	AC·	AA$^{A}_{G}$	AGU	CCA	UCA	CUU	AAU	GC·	GC·	AA$^{A}_{G}$
			A ↑					G ↓		
double mutant	AC·	AA$^{A}_{G}$	GUC	CAU	CAC	UUA	AUG	GC·	GC·	AA$^{A}_{G}$
	thr	lys	val	his	his	leu	met	ala	ala	lys

Figure 17 The base sequence of the wild type and the double mutant of lysozyme mRNA in phage T4.

6.4.5 The punctuation codons

It was known that the DNA molecule is many times longer than the average polypeptide, so the question arose as to how the DNA molecule is transcribed and translated into distinct polypeptides. In other words, it was necessary to discover the punctuation codons, or codons which would give the 'start' and 'stop' instructions for each polypeptide.

The initiation of polypeptide synthesis

Although Crick had demonstrated that the mRNA was translated from a fixed point, the nature of that point was not known until 1964. Sanger had isolated from *E. coli* a tRNA which was attached to a derivative of the amino acid methionine—

N-formylmethionine. This was very surprising because this derivative had a formyl group bonded to the amino end of the molecule. You may recall (S100[4]) that amino acids have the general formula:

$$\text{NH}_2\text{—}\overset{\displaystyle \overset{\text{H}}{|}}{\underset{\displaystyle \underset{\text{R}}{|}}{\text{C}}}\text{—}\overset{\displaystyle \overset{\text{O}}{\|}}{\text{C}}\text{—OH}$$

amino   carboxyl
group   group

R is dependent on the nature of the particular amino acid, and is different in each case.

In a polypeptide, amino acids are connected to each other by a bond, called a peptide bond, between the amino group of one and the carboxyl group of the next. Thus the general formula for a polypeptide is:

$$-\text{NH}_2\text{—}\overset{\text{H}}{\underset{\text{R}}{\text{C}}}\text{—}\overset{\text{O}}{\text{C}}\text{—}\overset{\text{R}}{\underset{\text{H}}{\text{N}}}\text{—}\overset{\text{H}}{\underset{\text{O}}{\text{C}}}\text{—}\overset{\text{H}}{\text{C}}\text{—}\overset{\text{H}}{\text{N}}\text{—}\overset{\text{O}}{\underset{\text{R}}{\text{C}}}\text{—}\text{C}\text{—OH}$$

peptide   peptide
bond   bond

Now, the interest in finding a tRNA with *N*-formylmethionine attached was that as the amino group was blocked by the formyl group it could not be used to form a peptide bond with the carboxyl group of another amino acid at that end of the molecule. Amino acids could be added only to the carboxyl end of *N*-formylmethionine.

Thus *N*-formylmethionine would not be able to occur in the middle of a polypeptide; it could only occur at one end of it. Streisinger's work with the T4 lysozyme had shown that it would have to be at the beginning of the polypeptide, as amino acids were connected to each other with their amino groups towards the beginning of the polypeptide and their carboxyl groups pointing towards the end.

The tRNA to which *N*-formylmethionine was connected contained the anticodon UAC; the codon for *N*-formylmethionine was therefore, AUG.

QUESTION Which amino acid had been assigned the codon AUG?

ANSWER Methionine. It is its only codon.

It turned out that in *E. coli* cells there were two types of tRNA that could become attached to methionine. Both contained the same anticodon. However, when methionine was attached to one sort of tRNA it was promptly formylated by an enzyme present in the system, resulting in a tRNA carrying *N*-formylmethionine.

Sanger proposed, therefore, that every polypeptide started with an *N*-formyl-methionine, and thus translation of the mRNA always started with the codon AUG. However, it was known that not all the polypeptides isolated from *E. coli* started with *N*-formylmethionine. This was resolved when it was discovered that in an *in vitro* protein-synthesizing system the newly formed polypeptides did have *N*-formyl-methionine as their first amino acid. It thus seems likely that the *N*-formylmethionine is removed after the formation of the complete protein.

If the AUG sequence does appear in the middle of a polypeptide sequence then methionine will be added to the growing polypeptide chain— *N*-formylmethionine cannot be introduced as it does not have a free amino group to join to the carboxyl group of the previous amino acid.

Recently it has been discovered that in some organisms codons other than AUG can initiate polypeptide synthesis, but these instances are probably relatively uncommon.

The termination of polypeptide synthesis

The discovery of the 'stop' signal at the end of the mRNA also involved the use of Benzer's *r*II system; *r*II mutants, you will recall, can grow on *E. coli* B strains but are unable to form plaques on *E. coli* K. However, in 1960, Benzer discovered that some of his *r*II mutants *could* grow on some strains of *E. coli* K but not on others; he called them *ambivalent mutants* and referred to the *E. coli* K which supported growth as *permissive* K *strains*, and those that did not, as *non-permissive* K *strains*.

Among Benzer's other *r*II mutants was one that had a deletion right across the boundary between the *A* and the *B* genes covering a third of *A* and a small part of *B*. When it was translated, a polypeptide was produced which in a complementation test was unable to provide the *A* function, but was able to provide the *B* function.

Now, as we have seen, the *A* and *B* genes normally function quite independently. Each presumably determines the synthesis of a particular polypeptide. However, nucleotide pairs at the junction of the *A* and the *B* genes are missing in the deletion mutant. Benzer surmised that this region must contain the initiation codons that normally signal the 'stop' of the *A* gene and the 'start' of the *B* gene, so that in this mutant, instead of two polypeptides being synthesized upon phage infection, only one would be formed. This polypeptide would be longer than either of the two normally produced, containing information specified in the beginning two thirds of the *A* gene plus the greater part of the *B* gene. As this polypeptide proved to have *B* activity, Benzer deduced that the small deleted part of the *B* gene must be relatively unimportant for gene *B* function.

Now let us see how Benzer used this deletion mutant to work out the existence of the 'stop' codon. First, he crossed ordinary *r*II*A* point mutants with the deletion mutant to find out what effect this had on the *B* gene function. In all cases, the *B* gene function remained intact in the double mutant (Fig. 18).

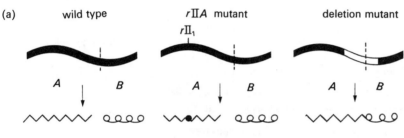

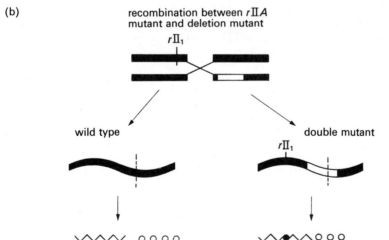

Figure 18 (a) A diagrammatic representation of the *r*II region in wild-type and mutant T4 phases and the polypeptides synthesized. (b) A cross-over between the *r*II*A* mutant and the deletion mutant gives a wild-type recombinant and a double mutant that retains the function of gene *B*.

When he crossed the deletion mutant to an ambivalent *r*II*A* mutant, however, the *B* function disappeared. He concluded from this result that the ambivalent mutant must have contained a type of mutation that stopped the synthesis of the polypeptide product. Thus the polypeptide could be synthesized quite normally up to the site of the *r*II*A* ambivalent mutation and then was halted, no part of the *r*II*B* gene being translated (see Fig. 19, overleaf).

From this evidence Benzer postulated the existence of *nonsense* codons which did not code for any amino acid. The *r*II*A* ambivalent mutant most probably carried a nonsense mutation (one resulting in a nonsense codon in the middle of the mRNA) and this led to the premature termination of the polypeptide chain.

Benzer next suggested that the *r*II ambivalent mutants were able to grow on some (the permissive) strains of *E. coli* K because these bacteria were able to read the

ambivalent mutant
permissive strain
non-permissive strain

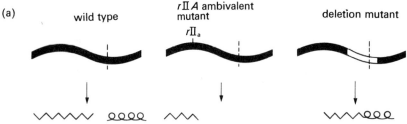

(a)

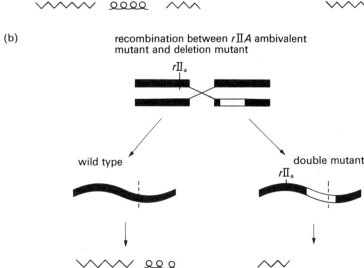

(b)

Figure 19 (a) A diagrammatic representation of the *r*II region of wild-type and mutant T4 phages and the polypeptides synthesized. (b) A cross-over between the *r*II*A* ambivalent mutant and the deletion mutant gives a wild-type recombinant and a double mutant with no gene *B* function.

nonsense codon as sense, and inserted an amino acid at that point in the polypeptide. That is, the permissive K strains contained *nonsense suppressors*. It was suggested that the nonsense suppressor could be a tRNA with a mutation in its anticodon that enabled it to read a codon that did not normally code for an amino acid.

nonsense suppression

Brenner applied this idea to another group of T4 mutants called the amber mutants, and was able to prove that nonsense codons did interrupt polypeptide synthesis.

Amber mutants ('amber' does not refer to a phenotype, it is just a pet name for these mutations in the laboratory in which they were discovered) of T4 have a mutation at some point in their genomes—a different point for each type of amber mutant— that prevents them from growing on normal strains of *E. coli*; they will grow only on permissive K strains.

QUESTION What is likely to be the nature of the amber mutations, given their host restriction?

ANSWER Permissive K strains are unusual in that according to Benzer they contain nonsense suppressors, that is, tRNAs that can 'read' nonsense codons and can thus prevent the premature termination of a polypeptide by inserting an amino acid at the point of the nonsense mutation. As amber mutants can grow only on these K strains the implication is that they contain nonsense mutations in essential genes, and can complete the lytic cycle only if nonsense suppressors are present in the cells they infect. The amber mutation can occur within any genes in the T4 genome.

Brenner had ten amber (*am*) mutants, all with a mutation in the region of the T4 genome which codes for the T4 head polypeptide. He first constructed a fine-structure genetic map showing the order of the *am* mutants within this region. He next digested the wild-type T4 head polypeptide with proteases (enzymes that break the peptide bonds at specific places in the polypeptide chain) and found that it could be broken down into eight fragments which could easily be separated by the technique of electrophoresis; the amino-acid sequence of each fragment could then be determined.

Brenner took each of his ten *am* mutants in turn and allowed it to infect a non-permissive strain of *E. coli*. Later in the phage growth-cycle he added radioactive amino acids to the growth medium. The *am* mutants infected the bacterial cells and began the synthesis of their own proteins using the protein-synthesizing machinery of their hosts. Phage assembly could not take place, but the proteins made were extracted from the bacterial cells. The protein extracts were then digested with specific enzymes and the fragments produced compared with the protein of the wild type. (Most of the protein made by phages in the later stages of their growth cycle is the head polypeptide, and the radioactive amino acids can be added at a particular time to make sure that they are nearly all incorporated into the head polypeptide.)

Using this procedure, Brenner was able to discover that each of his *am* mutants produced a different length of head polypeptide. This is illustrated in Figure 20 where the mutants have been arranged in order of increasing length of polypeptide produced. At the top of the figure the sequence of mutations in the head polypeptide gene is given. It is evident that the two sequences are the same, providing yet another proof for the colinearity of the DNA base sequence and the amino-acid sequence in protein. It can also be seen that the amber mutations do carry nonsense codons at various positions in the gene coding for the head polypeptide, resulting in the premature termination of the synthesis of the polypeptide.

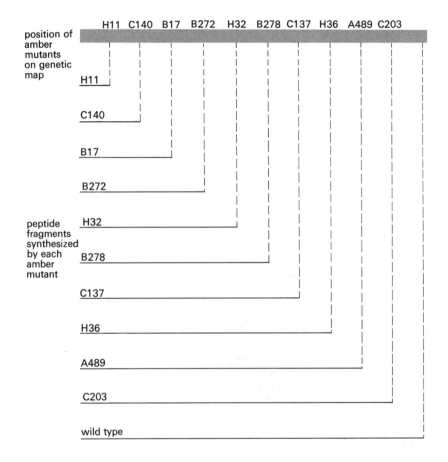

Figure 20 The genetic map of 10 amber mutants in the head-protein of T4 phage, and the corresponding peptide fragments produced by each mutant.

The next problem was to identify the nonsense codons. At this time (1965) very few codons had not been assigned an amino acid. UAA, UAG and UGA were the remaining codons available for assignment. Both Brenner and Garen, working independently on different systems, but using similar approaches, completed the identification of the nonsense or stop codons.

Garen worked with a nonsense mutation in *E. coli* that produced an incomplete phosphatase A polypeptide. He isolated a series of revertants to the wild type and then checked each revertant to see what amino acid was in the position where the nonsense codon had been. In the normal wild type, tryptophan would have been expected, but he found that his revertants contained a variety of amino acids at this point in their proteins (see Fig. 21, overleaf).

Given the codon assignments for each of these amino acids the nonsense codon must have been UAG, for only this codon could generate the amino acid found in each of the revertants with just one base change. Brenner who did similar experiments with the T4 *am* mutants came up with the same answer. UAG became known as the amber codon. Later work by both Brenner and Garen working independently on different systems showed UAA to be another nonsense codon, christened the 'ochre' codon. Finally Brenner and Crick identified UGA as the third nonsense codon.

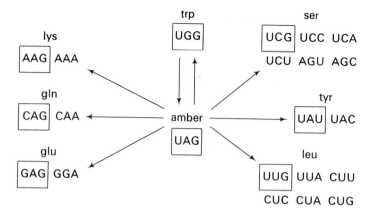

Figure 21 The amino-acid substitutions that accompany the reversions of a *pho*⁺ amber nonsense mutant. The codon assignments are indicated in boxes.

There is one final point worth mentioning. The discovery and assignment of nonsense codons was based on the use of mutants in which the mutation to a nonsense codon led to the premature termination of the polypeptide product. Are nonsense codons used by organisms to terminate their polypeptides normally? In other words, are they the same as the 'stop' codon at the end of the mRNA? It has now been proved by direct base-sequencing techniques (see Section 6.6) that they are, and furthermore, it is not uncommon to find two or more nonsense codons at the end of a mRNA— a sort of belt and braces approach!

6.4.6   The code *in vivo*

We still have to consider whether the details of the code demonstrated by the *in vitro* experiments we have discussed are true *in vivo*, and whether they apply to higher organisms. All the work we have considered so far was carried out with bacteria and phage.

Both Streisinger's experiments with the lysozyme of T4 and Garen's work with mutants of phosphatase A polypeptide were carried out in living organisms, although the interpretation does depend on the codons assigned to the amino acids determined *in vitro*. However, base-sequencing studies on a phage called R17 have shown that AUG really is the initiation codon, and this has also been shown to be true of yeast and mammals. We have already mentioned that the nonsense codons are used as termination codons *in vivo*; however, the actual termination process may be quite complicated.

The universality of the code can be considered at several levels. For example, the observations on the substitution of amino acids in mutants of organisms as widely disparate as tobacco mosaic virus, yeast and humans are compatible with the code as established for *E. coli*. At the *in vitro* level, extracts of *E. coli* containing all the requirements for transcription and translation can quite happily synthesize polio virus protein. Purified haemoglobin mRNA from rabbit cells can be translated in frog cells to give rabbit haemoglobin. The conclusion is that the code must be universal.

6.4.7   Summary of Section 6.4

1   The genetic code is defined as the relationship between the sequence of the four bases in the nucleic acid of genes and the sequence of amino acids in the proteins whose structure and nature they determine.

2   On the basis of theoretical prediction and the results of genetic analysis using suppressors of an acridine-induced *r*II mutant of T4 phage, the triplet nature of the code was established. Thus, sequences of three bases (called codons) code for each of the 20 different amino acids commonly found in proteins.

3   Analysis of the results of *in vitro* peptide synthesis using synthetic polynucleotides permitted the assignment of specific amino acids to 61 of the 64 triplet codons that can be formed from 4 bases.

4   The degeneracy of the code was clear from this codon assignment because most amino acids are coded for by more than one codon (sometimes by as many as four or even six codons).

5 Comparisons have been made between fine-structure maps of genes and the sequences of amino-acid changes in the related protein of each mutant mapped. They reveal a strict colinearity between the maps and the sequences and confirm that the code is read in one direction.

6 From *in vitro* studies of protein synthesis, the codon AUG was identified as the one that initiates protein synthesis.

7 Missense mutants arise as a result of a change within a codon that causes one amino acid to be inserted for another. The mutant protein is synthesized normally and differs from the protein of the wild type only by this change in a single amino acid.

8 Nonsense mutants arise from mutations that lead to the formation of a nonsense codon that terminates the polypeptide prematurely.

9 Studies of the suppression and reversion of nonsense mutants confirm that codons UAG, UAA and UGA cause the termination of the polypeptide chain.

10 Mutant tRNA molecules can bring about suppression of both missense and nonsense mutants. They recognize the mutant codons and insert the original (wild-type) amino acid in missense suppression and an amino acid for the termination signal in nonsense suppression.

11 The genetic code, as established by *in vitro* procedures using extracts of *E. coli*, applies to other bacteria, viruses, fungi and mammals as well.

Now try SAQs 6, 7 and 8 on p. 286i.

6.5 The control of gene expression

Once the molecular identity of the gene was elucidated, the next problem to solve was the control of gene expression. The living organism is not just a collection of genes; the action of genes must somehow be coordinated for the orderly progression of differentiation and development processes which translate genotype into phenotype.

Many genes are expressed biochemically as enzyme activity. At the elementary level, therefore, the study of the control of enzyme biosynthesis (and degradation) is the study of the control of gene expression. In this Section we shall discuss the biochemical and genetic evidence that was adduced in support of the existence of simple genetic systems regulating the biosynthesis of enzymes in *E. coli*. We shall then briefly discuss the control of gene expression in higher organisms.

6.5.1 Enzyme induction: the lactose system

E. coli cells contain the genes for the enzymes needed to utilize many substances as sources of carbon and energy. Ordinarily these genes are not expressed; the organism makes the enzymes required for the utilization of a particular compound only when that compound (or its analogue) is present in the medium. This is the phenomenon of enzyme *induction*, which enables the organism to physiologically adapt to the presence of a new source of nutrient in its environment. Although enzyme induction was discovered at the turn of the century, it was not until the late 1940s that Monod began to study its genetics. Most of the fundamental characteristics of enzyme induction have been worked out in the study of the lactose system in *E. coli*, and we shall concentrate on this system.

inducer

E. coli cells growing in a minimal medium with glucose as the carbon source contain very low levels of β-galactosidase. This enzyme catalyses the hydrolysis of β-galactosides such as lactose, so that the resultant simple sugars can be utilized for growth and reproduction. When the glucose in the medium is replaced by lactose, β-galactosidase activity starts to increase in the *E. coli* cell until within minutes the fully induced level is reached. Expressed as enzyme units per cell, the induced activity may be 1 000 times the uninduced activity.

Figure 22 shows an experiment in which the increase in the amount of β-galactosidase protein in the culture was expressed as a function of the increase in bacterial protein due to growth and reproduction. The inducer was added to a growing culture of *E. coli* so that the total amount of enzyme in the culture increased as the cells multiplied.

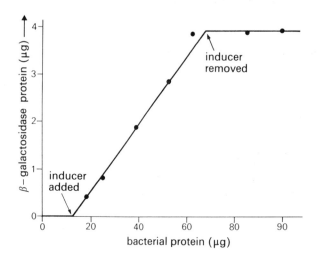

Figure 22 The induction of β-galactosidase in *E. coli*. The amount of β-galactosidase protein is plotted against the total bacterial protein.

QUESTION (a) What was the effect of adding the inducer?

(b) What was the relationship between the increase in β-galactosidase protein and the increase in bacterial protein after the addition of inducer?

(c) What was the effect of removing the inducer?

(d) In the induced state, what fraction of the bacterial protein was accounted for by β-galactosidase protein?

ANSWER (a) The effect of adding inducer was an almost immediate increase in β-galactosidase activity.

(b) The increase in β-galactosidase protein and the increase in bacterial protein were linearly related, that is, a constant rate of enzyme synthesis was maintained in the bacterial cells in the induced state.

(c) The effect of removing the inducer was an immediate cessation of β-galactosidase synthesis even though the bacterial protein continued to increase. That is, growth continued but no new β-galactosidase was synthesized.

(d) The slope of the curve between the addition and the removal of the inducer was about 0.066. This value meant that approximately 6.6 per cent of the total protein synthesized in the induced state was β-galactosidase protein.

How could we tell that enzyme induction involved the synthesis of new protein and not the activation of previously existing enzyme precursors? An experiment was carried out in which *E. coli* cells were grown for several generations in medium containing radioactive sulphur but without β-galactosidase inducer. (Sulphur is incorporated into sulphur-containing amino acids such as cysteine and cystine, which are in turn incorporated into protein.) The *E. coli* cells were then washed and transferred to a non-radioactive medium with inducer added. When β-galactosidase was fully induced, the enzyme was isolated and purified to check for the presence of radioactivity.

ITQ 9 What was the result observed that indicated that enzyme induction involved the synthesis of new enzyme molecules?

It turned out that other β-galactosides, some of which were not hydrolysed by β-galactosidase, could act as inducers. One example is *iso*propylthiogalactoside (IPTG), a potent inducer which is not broken down by the enzyme. It is therefore not utilized by the cell, and is particularly useful for kinetic studies as its concentration remains constant during an experiment.

The possession of β-galactosidase was not sufficient to ensure the utilization of lactose by *E. coli* cells. Another enzyme, β-galactoside permease, must be present to enable lactose to enter the intact cells which normally exclude the complex sugar.

It was shown that permease, and a third enzyme of unknown function, and acetylase, also had the same induction kinetics as β-galactosidase, and responded to the same specific inducers. In other words, all three enzymes were inducible specifically by the same inducers at more or less the same time after the addition of inducer; when the inducer was removed, the synthesis of all three enzymes ceased at about the same time. These enzymes are said to be *coordinately expressed*.

6.5.2 The *lac* operon

Jacob and Monod made numerous biochemical and genetic observations that enabled them to formulate in 1961 a hypothetical system of gene regulation. Since then, many of their predictions concerning the existence and mode of action of biochemical intermediaries have been verified.

This system essentially controls gene expression at the level of transcription. It is referred to as an *operon* and is now known to consist of the following components: **operon**

1   *Structural genes*, coding for enzyme proteins.   **structural gene**

2   An *operator gene*, contiguous to the structural genes and controlling their transcription.   **operator gene**

3   A *regulator gene*, coding for a *regulator* protein which binds to the operator.   **regulator gene**

4   A *promoter gene*, contiguous to the operator gene, and the site of attachment of mRNA polymerase (the enzyme that transcribes the DNA base sequence into mRNA).   **promotor gene**

5   *Effectors* which specifically bind to the regulator protein to alter its affinity for the operator.   **effector**

The *lac* operon is responsible for controlling the synthesis of the enzymes involved in the lactose system (see Fig. 23).

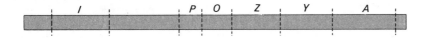

Figure 23 The genetic map of the *lac* operon. I = regulator gene, P = promoter gene, O = operator gene, Z, Y, A = structural genes for β-galactosidase, permease and acetylase, respectively.

The structural genes, *lazZ*, *lacY*, *lacA* (coding for β-galactosidase, permease, and acetylase, respectively) are in a cluster on the bacterial chromosome. The operator gene, *lacO*, is contiguous to the *lacZ* gene; the promoter gene *lacP* is contiguous to *lacO*. The regulator gene, *lacI*, on the other hand, is not adjacent to the other genes in the operon.

The regulator gene product in the *lac* operon is the *repressor* protein which binds to the operator to stop (or repress) the transcription of the structural genes by mRNA polymerase, which is attached to the promoter gene (Fig. 24(a)).   **repressor**

When an effector, or in the case of the *lac* operon, an inducer is added, it binds to the repressor protein, thereby preventing it from binding to the operator. When this occurs, the mRNA polymerase can move across, transcription proceeds and enzymes are synthesized (Figs. 24(b) and (c)).

6.5.3   Genetic evidence for the *lac* operon

We shall now summarize some of the genetic evidence for the existence and mechanism of action of the *lac* operon.

Evidence for the existence of contiguous structural genes

1   Missense mutations existed that affected each enzyme independently of the other enzymes.

2   The genes for the three enzymes mapped contiguously on the bacterial chromosome in the order *lacZ–lacY–lacA*.

3   Certain mutants mapping within the region of the structural genes caused the disappearance of more than one enzyme. These were called *polar mutants*. For   **polar mutant**

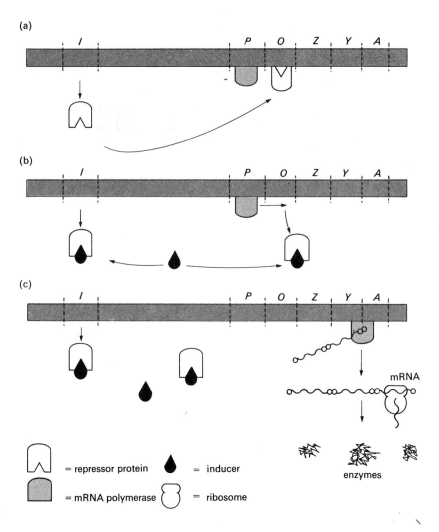

(a)

(b)

(c)

= repressor protein   = inducer

= mRNA polymerase   = ribosome

mRNA

enzymes

Figure 24 The mechanism of action of the *lac* operon. (a) When the inducer is absent, the repressor protein binds to the operator and no transcription of the structural genes takes place. (b) In the presence of the inducer, the repressor protein complexes with the inducer and comes off the operator. (c) Transcription of the structural genes takes place, followed by translation to give all three enzymes.

instance, a polar mutant within *lacZ* caused the disappearance of all three enzymes, whereas one occurring within *lacY* caused both the permease and the acetylase to disappear, leaving β-galactosidase unaffected. We now know that these were nonsense mutations that led to premature termination of the chain.

These observations supported the idea that the three enzymes were different proteins although they were transcribed as a unit and translated together. This forms the basis of the coordinate expression of the enzymes.

Evidence for the existence of a regulator gene

Two kinds of relevant mutant were discovered:

1 Mutants in which high levels of β-galactosidase, permease and acetylase were produced in the absence of inducer. These were called *constitutive mutants* and labelled *lacI⁻*. These mutants produced a defective repressor protein which could not bind to the operator to stop transcription of the structural genes; this resulted in continuous or constitutive synthesis of enzymes.

constitutive mutant

2 Mutants that had low levels of all three enzymes that could not be induced. These were termed *non-inducible mutants* and labelled *lacIˢ*. These mutants produced a repressor protein that had lost its affinity for the inducer, and therefore remained permanently bound to the operator even in the presence of the inducer.

non-inducible mutant

All *lacI* mutants mapped to the left of the structural genes but not adjacent to them.

Evidence for the existence of an operator gene

Other mutants were discovered that had the same phenotype as *lacI⁻* (they were constitutive for the synthesis of the three enzymes) but mapped at a different location. In fact, they mapped directly to the side of the *lacZ* gene away from *lacY* and *lacA*. These were termed operator-constitutive mutants (*lacOᶜ*). These mutants had a defect in the operator gene which reduced the binding of repressor protein.

Evidence for the existence of a promoter gene

A further group of mutants possessed the following properties:

1   They produced lower levels of all *lac* enzymes irrespective of whether the strain of *E. coli* was *lacI⁻* or *lacI⁺*, *lacO^c* or *lacO⁺*.

2   They mapped to the left of and adjacent to *lacO*. The locus was termed *lacP*, for promoter gene, because its function seemed to be that of promoting the transcription of the enzymes independently of all the other genes. *lacP* mutants were defective in the promoter region so that mRNA polymerase no longer attached itself effectively to this region.

Complementation and gene interaction in the lac *operon*

Much of the genetic evidence for the way the different genes interact came from complementation studies using partial diploidy. The system utilized is that of the F′ factor in *E. coli*. The F′ plasmid regularly carries with it parts of the genome of the host (see Unit 4, Section 4.6.2). When conjugation takes place, the F′ plasmid, in which some of the *lac* genes of the donor have been packaged, will be transferred to the recipient. The recipient strain then becomes diploid in parts of *lac* genome.

Interaction between structural genes and the operator gene

Table 6 shows the enzyme-induction characteristics of haploid strains and partial diploids of *E. coli* constructed with mutants in *lacZ* and *lacO*.

Table 6   Enzyme induction in haploid and partial diploids of *E. coli* carrying *lacO* and *lacZ* mutations

	Relative activity of β-galactosidase	
lac genotype	induced	uninduced
O^+Z^+	100	0.1
O^+Z^-	0	0.0
O^cZ^+	95	25.0
$O^+Z^-/F'O^cZ^+$	220	70.0
$O^+Z^+/F'O^cZ^-$	90	0.1

Compare the results obtained with the two partial diploids both of which are double heterozygotes for the mutations in *lacO* and *lacZ*. In $O^+Z^-/F'O^cZ^+$, the wild-type alleles were not on the same chromosome. No complementation occurred, as the synthesis of the enzyme was still constitutive (the uninduced enzyme activity was high). This result indicated that there is something unusual about the operator gene that distinguishes it from other genes. This idea is reinforced by the fact that no amber mutants have ever been isolated for the operator gene, suggesting that the *lacO* gene does not have a diffusible gene product.

Interaction between the regulator gene and the structural gene

Table 7 shows the enzyme induction characteristics of *E. coli* haploids and partial diploids carrying mutations in *lacZ* and *lacI*.

Table 7   Enzyme induction in haploids and partial diploids of *E. coli* carrying mutations in *lacZ* and *lacI*

	Relative activity of β-galactosidase	
lac genotype	induced	uninduced
I^+Z^+	100	0.1
I^-Z^+	100	100.0
$I^-Z^+/F'I^+Z^-$	240	1.0
$I^+Z^+/F'I^-Z^-$	240	1.0

Once again, compare the two partial diploids, and then answer the following ITQ.

ITQ 10 Does the I^+Z^- genotype carried by the F′ factor complement the I^-Z^+ genotype of the recipient bacteria?

It can be seen that the *lacI* gene regulates enzyme synthesis in a way quite different from *lacO*. The presence of complementation in the partial diploid between mutants indicates that the *lacI* gene gives rise to a diffusible gene product. This was predicted by Jacob and Monod some years before this gene product, the repressor protein, was actually isolated and characterized.

Interaction between the operator gene and the regulator gene

Experiments similar to those described above showed that various interactions occur between the operator gene and the regulator gene which supported the idea that the regulator-gene product reacts with the operator gene directly to control the transcription of the structural gene.

ITQ 11 What are the enzyme-induction characteristics of the following haploids and partial diploids?

(a) $I^+O^+Z^+$

(b) $I^sO^+Z^+$

(c) $I^+O^cZ^+$

(d) $I^sO^+Z^+/F'I^+O^+Z^+$

(e) $I^sO^+Z^+/F'I^+O^cZ^+$

Interaction of the promoter gene with the structural gene

Promoter mutants are similar to *lacO* mutants in that the partial diploids $P^+Z^-/F'P^-Z^+$ did not show complementation, indicating that the promoter had no diffusible gene product.

ITQ 12 How can we distinguish P mutants from

(a) *lacZ*, *lacY* or *lacA* mutants

(b) *lacI*⁻ or *lacI*ˢ mutants

(c) *lacO*ᶜ mutants?

Let us recapitulate the mechanisms by which the *lac* operon regulates enzyme biosynthesis (see Fig. 24). The primary signal to which the *lac* operon responds is the presence or absence of the inducer; the inducer is normally lactose. When lactose is absent, the *lac* operon is turned off: the repressor protein (the gene product of *lacI* which is synthesized continuously) binds to the operator, preventing the transcription of the structural genes by mRNA polymerase, itself attached to the promoter. No enzymes are synthesized in this state. As soon as the inducer is added to the cells, a complex forms between the repressor protein and the inducer, causing the repressor to come off the operator. This turns the *lac* operon on: the mRNA polymerase moves across the vacated operator and transcribes the three structural genes as a polygenic message (a messenger RNA containing the transcript of many genes) which is translated to give the *lac* enzymes.

Thus, all three enzymes are induced together in the presence of lactose. When the inducer is removed, or, in the case of lactose, is used up by the cells, the *lac* operon is switched off, once more by the binding of repressor protein to the operator, and the synthesis of all three enzymes stops.

Systems such as the *lac* operon are of great importance to cellular economy as they ensure that enzymes are synthesized only as they are required. When the enzymes are not required, the energy involved in synthesis could be redirected for other purposes.

We have presented a very simplistic view of the workings of the *lac* operon. Actually, the system is much more elaborate, and the mode of regulation is more finely attuned to the metabolic states of the cell. For example, the extent of binding of repressor protein to the operator depends on the concentration of the inducer present; this in turn determines the precise rate of transcription. The extent of binding of the mRNA polymerase to the promoter is under the control of one of the polypeptides which make up the enzyme molecule itself; this again influences the rate of transcription and hence the rate of production of the *lac* enzymes. Another protein is now known to bind to the promoter to control the rate of transcription. This is the cyclic-AMP receptor protein (CRP) which complexes with cyclic-AMP (the cyclic nucleotide, adenosine 3′,5′-monophosphate). The cyclic-AMP–CRP complex then binds to the promoter to initiate transcription.

The major feature to emerge from the study of the mechanisms of regulation is the specific binding of proteins to different regions of the DNA. This includes the binding of repressor protein to the operator, the mRNA polymerase and cyclic-AMP receptor protein to the promoter, for example. These specific protein–DNA interactions are involved in many other control mechanisms which have been discovered. In some instances, the structural enzymes themselves can take on such a regulatory role.

6.5.4 Other control systems

Many other operon-like systems exist in bacteria. The enzymes in an operon are characteristically those of a single biochemical pathway involved in the synthesis or breakdown of a certain metabolite. The *lac* operon we have discussed above is an example of a *negative control* system; the regulator is a repressor which binds to the operator to switch genes *off*. Other control systems are under *positive control*. In these systems, the regulator binds to the operator to switch genes *on*. An example of a positive control is found in the breakdown of the sugar arabinose *in E. coli*.

negative control
positive control

Enzymes are not always induced by the presence of specific metabolites: they may be repressed. In these cases, the enzymes are normally synthesized in the absence of the effector, but are switched off in its presence. Enzyme repression is characteristic of enzymes involved in biosynthetic pathways leading to an essential metabolite, for example, the enzymes involved in the biosynthesis of histidine. In this instance, histidine, the end-product of the pathway acts as the effector. The addition of histidine to the growth medium immediately switches off the synthesis of the enzymes. The mechanism of enzyme repression involves the production of a co-repressor protein by the regulator gene. The co-repressor will not bind to the operator except in the presence of the effector. Therefore, transcription occurs when histidine is absent and stops when histidine is added.

Finally, the clustering of structural genes does not occur in all control systems. The operons of many metabolic pathways have structural genes which are scattered singly or in small groups throughout the genome. In these cases, each small group, or single structural gene, will have its own contiguous operator and promoter.

> **ITQ 13** Why is it necessary to have an operator contiguous to each singly occurring structural gene or each scattered group of structural genes?

6.5.5 The control of gene expression in higher organisms

The sort of control systems described are responsible for reversible changes in gene transcription in response to environmental changes. In addition to the necessity for these reversible and short-term adaptive changes in enzyme activity, higher organisms, to varying degrees, must undergo the sort of permanent changes in gene transcription associated with differentiation and development. The genetic basis of the permanent changes is far from being understood (see Unit 8) although much effort has been directed towards elucidating the molecular basis of differentiation and development in recent years.

There are some instances of reversible and adaptive changes in gene expression in higher organisms. Many liver enzymes of mammals can be induced; for example, in humans, liver alcohol dehydrogenase can be induced by alcohol in the diet.

Hormones play a special role as effectors, inducing or repressing different enzymes in special target cells. They are of special importance during development. In insects, the hormone ecdysone triggers moulting. When ecdysone is added to insect larvae, it stimulates chromosome puffing, the pattern of which is similar to those seen in normal moulting (Unit 8, Section 8.6.3). These puffs are believed to represent sites of active gene transcription.

So far, we have described the control of gene expression at the level of transcription. Other forms of control are possible. It is known for example, that mRNAs in higher organisms are long-lived; their half-lives are estimated to be of the order of hours or even days. As the efficacy of control at the level of transcription depends crucially on the mRNA being short-lived (as in bacteria), control in higher organisms, particularly as regards short-term, reversible, enzyme changes, may well involve the regulation of the rate of translation or the rate of protein degradation. Some evidence for such regulation exists.

Genetic components of a similar nature to those in the bacterial operon, such as effectors, regulator genes and operator genes, have been tentatively implicated in higher organisms, but the picture is much more complex. This is not surprising as the processes and structures involved are far more complex than those of bacteria. Some workers, like Britten and Davidson (1969), have postulated the existence in higher organisms of hierarchies of genes controlling differentiation; this is essentially an elaboration of the simple operon-like model with positive control.

Certain regions of the genome, called sensor genes, bind specific effectors, resulting in the transcription of a number of regulator genes to give a family of regulator molecules. These in turn bind to specific operators scattered throughout the genome contiguous to different structural genes to switch on transcription. Thus, depending on the sensor activated, a different battery of structural genes will be transcribed (and therefore translated). These structural genes will not necessarily be related to each other, as for example, are genes coding for enzymes of a particular metabolic pathway. The sensor genes can overlap in the family of regulator genes they control so that the possible combinations of enzymes and proteins transcribed are multiplied enormously. It is reasoned that regulation in higher organisms depends on the multiplicity of control elements rather than on the multiplicity of structural genes.

ITQ 14 Figure 25 is a schematic representation of a hypothetical gene-control system in higher organisms. S_1, S_2, and S_3 are sensor genes: I_A, I_B, and I_C are regulator genes: O_A, O_B, and O_C are the corresponding operator genes recognizing the three regulator genes, and A, B, and C are corresponding structural genes, the synthesis of which is being controlled.

(a) Which polypeptides are synthesized when sensor S_1 is activated by the appropriate effector?

(b) Which polypeptide is common to all three batteries of genes? (Each battery is controlled by one sensor.)

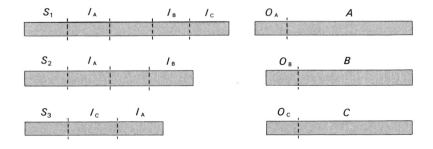

Figure 25 A schematic representation of a gene control system in eukaryotes.

Candidates for the specific effectors that activate sensor genes include the hormones and their receptor-proteins. The special regulator genes that coordinate the expression of unrelated proteins have been identified in simple eukaryotes such as the fungus *Aspergillus*.

6.5.6 Summary of Section 6.5

1 The control of gene expression is studied in the genetic regulation of enzyme biosynthesis.

2   The synthesis of certain enzymes is induced by the presence of their specific substrates or substrate analogues. Further investigations show that the induction is immediate, and that other enzymes involved in the metabolic pathway of the inducer are induced at the same time.

3   The enzymes involved in the breakdown of lactose in *E. coli* are synthesized in the presence of lactose. The system of control is that of an operon which consists of the following elements:

(a) structural genes—coding for the three enzymes involved in lactose breakdown. These are adjacent to each other and are transcribed and translated as a unit.

(b) an operator gene—contiguous to one end of the group of structural genes and controlling the transcription of the structural genes.

(c) a regulator gene—not adjacent to the group of structural genes; it codes for a regulator protein that regulates transcription of the structural genes by binding to the operator gene.

(d) a promoter gene—contiguous to the operator gene on the side opposite to that of the structural genes; it is the site at which mRNA polymerase attaches itself during the initiation of transcription.

(e) an effector—a small molecule that interacts with the regulator protein to alter its binding to the operator.

4   The *lac* operon is switched off when the regulator (or repressor) is bound to the operator, and switched on when the repressor is removed from the operator in the presence of the effector.

5   The *lac* operon is under negative control: the regulator switches genes off by binding to the operator. A system is under positive control if the regulator binds to the operator to switch genes on. Both positive and negative control systems exist in bacteria.

6   The control of gene expression in higher organisms may take place at the level of translation or the degradation of protein as well as at the level of transcription.

7   Elaborations of the operon model have been postulated for gene control in higher organisms. Higher organisms may depend on the multiplicity of control elements rather than the multiplicity of structural genes.

Now try SAQs 9 and 10 on p. 286i.

6.6   Conclusion and perspectives

In this Unit we have attempted to sketch the bare outlines of the exciting developments within the 20-odd years (from the 1940s to the 1960s) during which the key discoveries about the structure and function of the genetic material were made.

Part of the original intellectual impetus to study the physical basis of genetic information came from the publication in 1945 of the physicist Erwin Schrödinger's book, *What is Life?* (*HIST*, Section H.5.5). In this book, Schrödinger identified the primary problem of life as heredity. To what kind of molecular structure can one attribute the stability of the genetic information that is passed on almost unchanged from generation to generation? With remarkable insight, Schrödinger postulated that the genetic material must be a large 'aperiodic crystal' composed of a succession of a small number of elements, the exact nature of which constituted the 'hereditary code'.

The success of the search for the molecular mechanisms of gene action attests to the power of the analytical, reductionist approach in science (*HIST*, Section H.1). But are we any closer to providing a full explanation of the nature of living organisms and the distinction between living and non-living? Some would argue that life began with the emergence of self-replicating structures, such as DNA. But others would maintain that there was more to life than mere self-replication, and that the unfolding history of each individual living creature and the changing pattern of life through evolutionary time remain unaccounted for and unexplained.

But even the story of the gene itself is not complete. Within the last decade, a lot of details concerning the processes of DNA replication and recombination, transcription and translation have been worked out.

As this Unit goes to press, one of the long cherished rules of genetics has been violated. It appears that the one polypeptide–one gene relationship has been ignored by the bacteriophage φX174. This phage possesses at least one gene that codes for two separate polypeptides. The reading frame for the second polypeptide starts from somewhere in the middle of the gene and continues on to the end. With a phase-shift of one base, the reading of the second message from this gene gives a totally different polypeptide.

Meanwhile, these advances are beginning to make possible the application of new genetic techniques to the development of what has become known as 'genetic engineering' (you should see also Unit 15, Section 15.4.5). At the molecular level, this term refers to the production of new genes or combinations of genes by direct manipulation of DNA molecules, bypassing the normal process of genetic recombination and consequently overcoming the interspecific barriers to sexual reproduction that occur in nature. These methods open up the prospect of realizing an age-old dream, the creation in the laboratory of novel organisms with a known genetic constitution.

Molecular genetic engineering depends on three important techniques all developed within the last 5 years. The first technique is that of base sequencing, (developed by Sanger and his colleagues here in Cambridge) by which the base sequence of any stretch of DNA can be rapidly determined. The second technique is the construction of recombinant DNA *in vitro*. A combination of enzymes is used to cut DNA molecules in specific places and to splice them together again to create new molecules (recombinant DNA). This technique has enabled the insertion of genes from diverse sources into bacterial plasmids, so that the genes could be cloned or mass-produced as the plasmid replicates within the bacterial cell. (DNA base-sequencing work has been made a lot easier: the amount of the DNA to be sequenced can be increased enormously by cloning procedures.) One potential application of the technique of recombinant DNA is to insert human genes into the bacterial genome in order to synthesize medically valuable proteins, such as insulin, which at present have to be isolated from tissues in which they naturally occur. One obvious problem which would need to be overcome is that human genes are not transcribed in bacteria because their control regions, for example, promoters, are not recognized by the bacterial machinery. The third technique is the chemical synthesis of the DNA of any desired base sequence developed by Har Gobind Khorana and his colleagues at MIT, Cambridge, Massachusetts. This may eventually enable molecular geneticists to make a control sequence of bacterial origin that can be stuck next to a human structural gene, provided the required DNA sequence is known. Insertion of this complex into the bacterial genome should then result in transcription and translation of valuable human proteins. Another application of the techniques of molecular genetic engineering may be the creation of new strains of food crops by inserting into their genomes useful genes such as those involved in nitrogen fixation. These techniques have begun to be developed in the laboratory (see TV programme 16) but it remains to be seen whether, in practice, they will come to bypass traditional methods in plant and animal breeding (see Unit 12).

The existing tools of genetic engineering can also be used in probing the molecular and functional organization of the genomes of higher organisms. The problem of differentiation and development in higher organisms has challenged and intrigued biologists for many years. Molecular genetics has offered very few insights thus far. Indeed, some developmental biologists remain unconvinced that the analytical methods of molecular genetics are appropriate to the solution of this problem. The crux of the debate appears to be: can we completely describe differentiation and development in molecular terms or do we need also to consider, for example, the relationships between these molecules in space and time? (Unit 8 will give you some further insights into this discussion.)

Meanwhile, in the midst of the enthusiasm and excitement generated by the new technology of molecular genetic engineering, there came warnings of its potential hazards. These issues were first raised among laboratory workers who were using the new techniques and then they were discussed at a major conference of molecular biologists and geneticists in California in 1975. The warnings were rapidly taken up elsewhere: by government committees in both the USA and Britain, by technicians and others who feared exposure to the possible risks, and by concerned citizens in the vicinity of laboratories engaged in the work. As this Unit is being written, a complex public and legal battle is being fought by the Mayor and citizens of Cambridge, Massachusetts to prevent recombinant DNA experiments being carried out in laboratories within the city, and in many laboratories a moratorium on work in this area has been called until the issues are resolved.

The fear is that the techniques of making recombinant DNA can introduce foreign genes into otherwise harmless bacteria, which could then transform them into dangerous human pathogens. If these were to escape from the laboratory in which they were being studied, they could cause an epidemic of diseases new to medicine, whose control and cure would be a major problem. The chances of bacteria escaping, unless rigid control measures are maintained (for instance, the use of only 'crippled' strains of organism for recombinant work, etc.) cannot be discounted. It is true that this fear is not based on established scientific knowledge but rather on the notable *lack* of understanding of the possible consequences involved in the application of a rapidly growing and potentially very powerful technology. This does not make it any less real. In a way the issues are analogous to those faced by physicists and others involved in making the relevant political decisions about whether to go ahead with the development of nuclear power for the generation of electrical energy. Biologists themselves are divided. Some, such as Erwin Chargaff, one of the key figures in the elucidation of the chemical structure of DNA in the 1940s and 1950s, have argued that the possible gains in knowledge and technique that recombinant work may provide are miniscule in comparison with the risk of irreparable damage. Others, anxious to proceed with their research would maintain that the risks are no greater than those experienced fairly routinely in a normal pathology laboratory, and that the whole issue has been exaggerated because of the atmosphere of concern raised by other quite legitimate fears about environmental pollution and the conscious or unconscious development of biological warfare. What is certain is that the decisions should not be left just to the scientists who want to work in this area but must involve all those people likely to be affected.

Appendix 1 Protein synthesis

The sequence of events is:

1 Specific segments of DNA are transcribed into mRNA molecules in the nucleus.

2 mRNA and tRNA move from the nucleus into the cytoplasm.

3 Aminoacyl-tRNA is formed and the mRNA binds on to the ribosomes. The formation of aminoacyl-tRNA requires ATP.

4 Reactions take place as described in the translation sequence below.

5 The completed polypeptide chain is released from the RNA and the ribosomes. This probably requires ATP.

6 The polypeptides are folded into their final shape.

Translation

The process of translation is illustrated in Figure 26. First (Fig. 26(a)), mRNA binds to the smaller ribosomal subunit. Next (Fig. 26(b)), an aminoacyl-tRNA that has a group of bases complementary to a group in the messenger binds to both parts of the ribosome at site P. But in order to start the synthesis of a whole polypeptide chain, a tRNA carrying a special amino acid, *N*-formylmethionine, and having an anticodon complementary to AUG must bind to the ribosome at site P. The next tRNA molecule approaches the other binding site, A.

In Figure 26(c) the second aminoacyl-tRNA binds to the ribosome at site A, recognizing the next codon in the messenger. The next codon is GCC, and the tRNA that recognizes that codon always carries the amino acid, alanine.

Figure 26

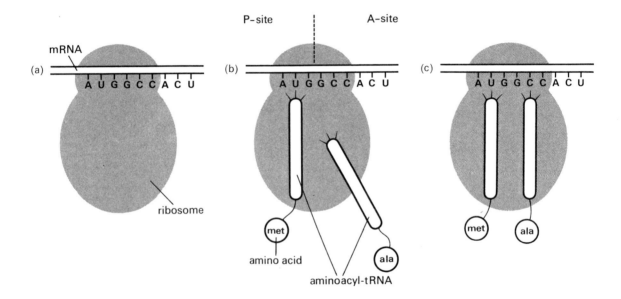

In (Fig. 26(d)) the two amino acids are now next to each other and a reaction takes place between them to form a 'peptide bond', the bond that connects amino acids to form a polypeptide or a protein chain. A reaction releases the tRNA from site P (Fig. 26(e)). The second tRNA, which now has both amino acids attached to it, moves from site A to site P. A third aminoacyl-tRNA, threonyl-tRNA, carrying the amino acid threonine, recognizes the next messenger group of bases, ACU, and attaches to site A, which is exposed by the movement of the mRNA relative to the ribosome. In Figure 26(f) the sequence is repeated as from stage 3.

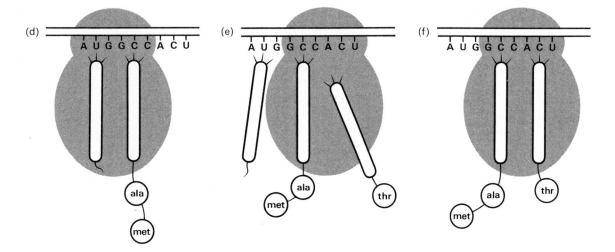

This sequence of events ends when one of the 'stop' codons occurs in the mRNA (Fig. 26(g)). No further amino acids are added to the polypeptide chain. Instead, both it and the tRNA separate from the ribosomes (Fig. 26(h)) and from each other. The polypeptide is now free in the cytoplasm of the cell.

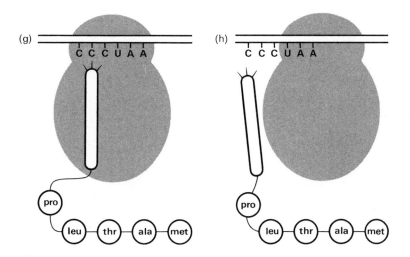

Self-assessment questions

SAQ 1 Three mutant auxotrophic strains (I to III) of *N. crassa* requiring methionine for growth were tested for their ability to grow on minimal medium supplemented with known metabolic precursors of methionine. The results are shown below:

		Growth response			
		minimal medium plus			unsupplemented minimal medium
Mutant	cysteine	cystathione	homocysteine	methionine	
I	−	−	−	+	−
II	−	+	+	+	−
III	−	−	+	+	−

(+ = growth; − = no growth)

From the data above, work out the likely sequence of the biochemical pathway leading to the synthesis of methionine, and indicate the site of the metabolic block in each mutant.

SAQ 2 The wild-type character can be restored by complementation or by recombination. Distinguish between these processes.

SAQ 3 Distinguish between a complementation map and a deletion map.

SAQ 4 The ends of 6-deletion mutants (C, D, E, F, G and H) define 10 separate divisions of the *r*II region of phage T4.

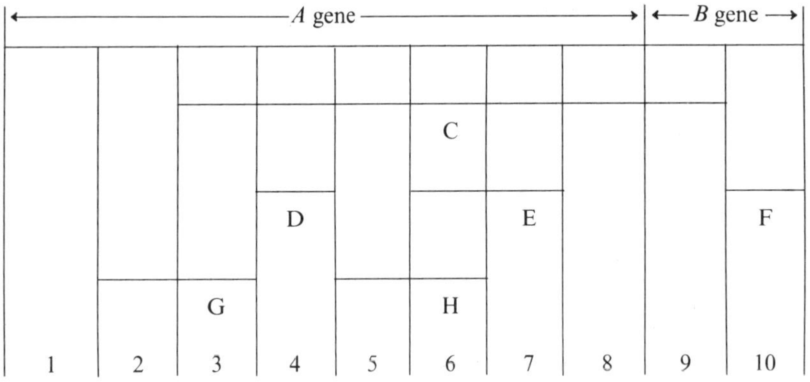

The results of complementation and recombination tests between four other *r*II mutants, K, L, M and N, and the deletion mutants were as follows:

	C	D	E	F	G	H
K	O	R	R	C	R	O
L	O	C	C	R	C	C
M	O	O	O	C	R	O
N	O	R	O	R	R	O

O = no lysis
C = complementation
R = recombination

(a) In which regions of the map can you place K, L, M and N?
(b) In what way do mutants M and N differ from K and L?

SAQ 5 Some *E. coli* auxotrophic mutants were tested for their reversion to the wild type in the presence of different mutagens. The following results were obtained.

		Reversion		
Mutants	no addition	base analogue	acridine orange	methylmethane-sulphonate
1	−	−	+	−
2	(−)	+	−	+
3	(−)	(−)	−	+

(+ = reversion;
(−) = almost no reversion;
− = no reversion)

What can you infer from the data concerning the molecular basis of the different mutations involved?

SAQ 6 As a result of single-step mutations the following amino-acid changes were observed at one particular site in an enzyme:

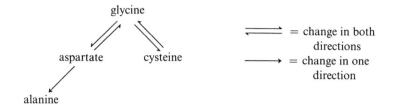

(a) Using the genetic code in Table 5 on p. 273, decide which codons would fit this set of events.

(b) Which mutant pair(s) could recombine to form the wild type?

(c) Which mutant pair(s) would not recombine to form the wild type?

SAQ 7 Which of the following statements are true?

(a) A single base substitution is a missense mutation.

(b) A missense mutation is one that results in a single amino-acid substitution.

(c) A deletion of base-pairs is a nonsense mutation.

(d) A nonsense mutation leads to premature chain-termination in the polypeptide.

(e) A suppressor mutation is a mutation that suppresses the expression of a mutant phenotype and occurs at a site distinct from the site of the mutant it suppresses.

SAQ 8 Match each of the DNA base changes (a)–(e) with one or more mutations (i)–(v) that it can give rise to.

DNA base changes

(a) A single base-pair substitution

(b) A deletion of one base pair

(c) A deletion of three adjacent base pairs

(d) An insertion of one base pair

(e) A deletion of a 'stop' codon, and a 'start' codon of the next gene.

Mutations

(i) A missense mutation involving the substitution of a single amino acid

(ii) A nonsense mutation involving a premature termination of the polypeptide chain

(iii) A mutation involving the absence of a single amino acid from the polypeptide chain

(iv) A frame-shift mutation in which all amino acids distal to the mutation are altered

(v) A giant polypeptide made up of the products of two contiguous genes

SAQ 9 Earlier in this Unit (Section 6.1) we saw that the one enzyme–one gene relationship was modified to the one polypeptide–one gene relationship. Can you think of any genes we have named so far that do not specify a polypeptide?

SAQ 10 You have almost reached the end of the Unit with the completion of Section 6.5. How would you now define a gene in molecular terms? Try to cover all the cases you have come across.

Answers to ITQs

ITQ 1 (*Objective 2*) Mutant 1 accumulates W, which does not support the growth mutants 2 or 3. This indicates that mutant 1 has a metabolic block after W and that mutants 2 and 3 also have their blocks after W. W is therefore, the first precursor. The pathway is

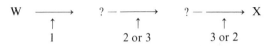

Mutant 2 accummulates Y, which supports the growth of 1 but not 3. This indicates that mutants 2 and 3 have their metabolic blocks after Y; but mutant 1 has its block before Y. The pathway is now

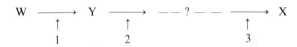

Mutant 3 accummulates Z, which supports the growth of both mutants 1 and 2. This indicates that both mutants 1 and 2 have their blocks before Z, but mutant 3 has its block after Z. The complete pathway is

ITQ 2 (*Objective 2*) Let us represent the normal enzyme molecule as (AB). The individual homozygous for the mutation in the *A* gene will have the enzyme (A*B)*, and the individual homozygous for the mutation in the *B* gene will have the enzyme (AB*). Both (A*B) and (AB*) are inactive.

Let *A* and *a* represent the normal and mutant alleles which specify polypeptides A and A* respectively; and *B* and *b* represent the normal and mutant alleles which specify polypeptides B and B* respectively. The inheritance of the alleles is as follows:

$$aaBB \times AAbb$$

$$AaBb$$

Four kinds of polypeptides will be synthesized in the offspring:

$$A, A*, B, B*$$

and four kinds of enzyme molecules will be formed in the proportion 1:1:1:1

$$(A*B*), (A*B), (AB*), (AB)$$

As only (AB) is active, the enzyme activity in the offspring will be about 25 per cent of normal. This is quite often sufficient to restore the phenotype to 'normal'. In other words, if the enzyme catalyzes a reaction in the synthesis of an essential metabolite, 25 per cent of the activity may be sufficient to allow enough of the metabolite to be synthesized for growth and development. Thus we say that complementation has occurred.

ITQ 3 (*Objective 3*) The mutation is in subdivision A1b2.

ITQ 4 (*Objective 3*) The results with deletion set 1 give an approximate location of the new mutant. As it fails to recombine with 1272, 1241 and J3, PT1 and PB242, but does recombine with A105 and 638, it must be in the region covered by PB242 and not by A105, that is, in region A5. A more precise location can be worked out from the result with deletion set 2. Failure to recombine with PB28 and P605, but recombination with 1589 and PB230, places this mutant within subdivision A5b.

ITQ 5 (*Objective 4*) The sequence of events leading to base-pair substitutions starting from the incorporation of 5-bromouracil (5BU) and 2-aminopurine (2AP), respectively, is shown in Figure 27. Both lead to a substitution in the A–T → G–C direction.

* *Note*: In (A*B) the asterisk represents a mutant polypeptide.

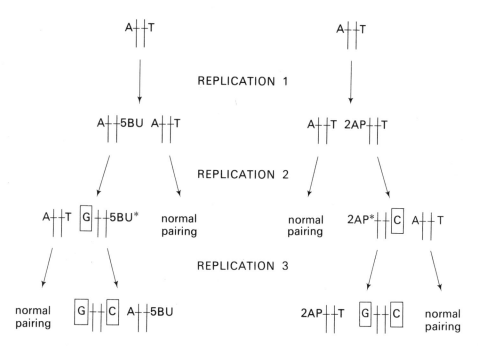

*tautomer of analogue inducing mispairing

Figure 27 The induction of base-pair substitutions by base analogues, 5-bromouracil (5-BU) and 2-aminopurine (2-AP).

ITQ 6 (*Objective 4*) Base analogues increase the reversion rates of mutants induced by base analogues. This can be predicted from their capability to induce two-way transitions, that is, A–T $\rightleftharpoons$ G–C.

ITQ 7 (*Objectives 5 and 6*) From Table 4, group 1, the possible codons for the amino acids valine, leucine, cysteine, tryptophan and glycine are UGU, UUG, GUU, GUG, GGU and UGG. From group 2, we know that of UGU and GUG one codes for cysteine and one for valine, so this narrows the field somewhat. Finally, the repeating trinucelotides of group 3 can be read as either UGU, UUG or GUU, and code for either leucine, cysteine or valine. This confirms that UGU codes for either cysteine or valine and suggests that UUG or GUU code for either valine or leucine. Thus, there is no firm indication of the amino-acid specificity of the four triplets UGU, GUG, UUG or GUU, but because between them they code for only three amino acids, some degeneracy is indicated. Clearly, going back to group 1, of GGU and UGG one codes for tryptophan and one for glycine.

ITQ 8 (*Objective 6*) The mutational changes in the strains were most likely single base substitutions. The sequence of changes is given in Figure 28. As there is only one codon for methionine, we can assign the codons for the other amino acids un-ambiguously. (The correct codons are enclosed in boxes.)

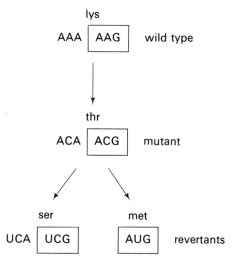

Figure 28 Amino-acid substitutions in the mutant and the revertants in an enzyme of *E. coli*. The correct codons are enclosed in boxes.

ITQ 9 (*Objectives 8, 10 and 11*) The β-galactosidase protein is not radioactive.

ITQ 10 (*Objectives 9–11*) Yes, there is complementation; the partial diploid is inducible.

ITQ 11 (*Objective 9*) (a) Wild type.

(b) Non-inducible. *lacI*ˢ produces a repressor that has lost its affinity for the inducer.

(c) Constitutive for the synthesis of β-galactosidase because of a defective operator.

(b) Non-inducible. *lacI*ˢ produces a repressor that has lost its affinity for the operator on both chromosomes (as it is diffusible) to stop transcription.

(e) Constitutive. The defective operator cannot bind any repressor protein effectively (produced by either *lacI* or *lacI*ˢ) so transcription of *lacZ* proceeds in the presence or absence of the inducer.

ITQ 12 (*Objective 10*) (a) *lacP* mutants do not specifically lack any one of the enzymes produced by *lacZ*, *lacY* or *lacA*, and do not map within any of the structural genes.

(b) *lacP* mutants, unlike *lacI*⁻, which is constitutive for the synthesis of all enzymes, will be similar to the *lacI*ˢ mutants in having a low level of enzyme activity both in the presence and absence of inducer. The difference between the induced and uninduced levels in *lacP* mutants will, however, be greater than that in the *lacI*ˢ mutant. *LacP* mutants, of course, map in a different location to *lacI*.

(c) *lacP* mutants are distinguished from the *lacO*ᶜ mutant in having lower levels of all enzymes, induced or uninduced. *LacP* mutants, once again, map in a different location.

ITQ 13 (*Objective 9*) The control of transcription of the structural genes depends on the existence of a site at which the regulator protein can bind. Therefore there must be an operator contiguous to each isolated structural gene or each group of structural genes within the operon.

ITQ 14 (*Objective 9*) (a) Polypeptides specified by structural genes *A*, *B* and *C*.

(b) The polypeptide of gene *A*.

Answers to SAQs

SAQ 1 (*Objective 2*) Mutant I cannot grow on any precursor, so its block is immediately before methionine. Mutant II can grow on all precursors except cysteine so the block occurs before cystathione and homocysteine but after cysteine. Mutant III can only grow on the precursor homocysteine, so its block occurs before homocysteine but after cysteine and cystathione.

Cysteine failed to support the growth of any mutant so its position in the metabolic pathway is before any step that is blocked. Cystathione supports the growth of one mutant only whereas homocysteine supports the growth of two mutants. The complete pathway and the sites of metabolic block are:

cysteine ⟶ cystathione ⟶ homocysteine ⟶ methionine

       ↑   ↑   ↑

      II   III   I

SAQ 2 (*Objective 3*) The restoration of a wild-type character by complementation is due to a functional cooperation between two different genomes defective in different genes.

The restoration of a wild-type character by recombination results from actual exchanges of DNA or chromosomal segments between two different genomes carrying mutations at different sites on the DNA or chromosome. In this process an actual wild-type gene or allele is reconstructed from combinations of DNA in the two mutant genomes.

SAQ 3 (*Objectives 1 and 3*) The essential differences between a complementation map and a deletion map are:

(a) A complementation map is constructed from the results of complementation tests which give information about the identity of the functional deficiencies in each of the two mutants involved in the test. A deletion map is constructed from results of genetic crosses which give information about the identity of the actual sites of mutation involved in the chromosome of the two mutant genomes in the cross.

(b) No recombination event occurs in complementation whereas recombination is always involved in crosses used to construct a deletion map.

(c) Mutants that do not complement each other may recombine, thus there is not necessarily any correspondence between a complementation map and a deletion map.

(d) The overlapping of lines in a complementation map signifies the lack of complementation between different mutants owing to a coincidence in their functional deficiency. The overlapping of lines in a deletion map signifies the lack of recombination owing to an overlap in the region of the genomes deleted in the different mutants.

SAQ 4 (*Objective 3*) (a) All four mutants fail to recombine with deletion C and so are 'covered' by it. The following additional deductions can be made: K complements F and is therefore in the *A* gene; it recombines with D, E and G but not with H, and so is located in region 5. L complements all deletions of gene *A* and so is located in the *B* gene; as it recombines with F it is located in region 9. M complements F and is therefore in the *A* gene; it recombines only with G and not with D, E or H, so it must be a deletion extending from sections 4–7 (or 8) inclusive. N recombines with D, G and F but not with E or H and so is a deletion from sections 5–9 inclusive, that is, it extends across the boundary between genes *A* and *B*.

(b) M and N are clearly deletions but, on the other hand, it cannot be said with certainty that K and H are point mutants; they could be small deletions in either section 5 or 9 respectively.

SAQ 5 (*Objective 4*) Mutant 1 is very likely a deletion mutant so it does not revert spontaneously but is reverted by acridine orange, which could reinsert missing base pairs. Mutant 2 is most probably a transition mutant so its spontaneous reversion rate is much enhanced in the presence of base analogues and methylmethanesuphonate, both of which induce transitions. Mutant 3 is most likely a transversion mutant as its spontaneous reversion rate is enhanced in the presence of methylmethanesulphonate, which is known to induce transversions (as well as transitions), but not in the presence of base analogues.

SAQ 6 (*Objective 6*) (a) The glycine codons are GGU, GGC, GGA or GGG, so any base could occupy the third position in the codon (degeneracy). GAU and GAC are the codons for aspartate, so a mutation at the second base (G → A transition) would result in the substitution of aspartate for glycine or vice versa. All four codons for alanine have GC as their first two bases, so an A → C transversion at the second base in the codon for aspartate would result in alanine being substituted for aspartate. Finally, the cysteine codons are UGU or UGC, so G → U transversions of the first base of the glycine codon would result in the substitution of cysteine for glycine.

In summary

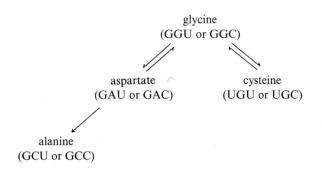

(b) The wild type is either GGU or GGC and recombination within the codon required the regeneration of the GG combination of the first two base pairs. Thus recombination between the aspartate and cysteine, and alanine and cysteine mutants could give the wild type.

(c) By the same token, the GG combination could not result from recombination between the aspartate and alanine mutants.

SAQ 7 (*Objective 7*) (a) and (c) are false, and (b), (d) and (e) are true.

(a) is false because a missense mutation merely describes the change in the polypeptide. At the level of the DNA, a single base-pair substitution can give rise to either a missense or a nonsense mutation depending on the resultant codon.

(b) is correct; the statement is a definition of a missense mutation.

(c) is incorrect because, once again, nonsense describes the change in the polypeptide. At the level of the DNA, a deletion may or may not give rise to a nonsense mutation.

(d) is true; the statement defines a nonsense mutation

(e) is true; the statement defines a suppressor mutation.

SAQ 8 (*Objective 7*) (a) (i), (ii);   (b) (iv), (ii);   (c) (iii);   (d) (iv), (ii);   (e) (v).

SAQ 9 (*Objective 11*) The promoter gene and the operator gene do not specify any polypeptide. They are not even transcribed. The genes coding for tRNAs are transcribed but not translated into polypeptides. Lastly, the genes coding for ribosomal RNA are also transcribed but not translated.

SAQ 10 (*Objective 11*) This is very difficult. In fact, few geneticists would try to define a gene, and not only from fear that the definition will become obsolete. It could be something along these lines: 'A gene is a stretch of the DNA molecule that has a well-defined function'. You can see that this includes promoter genes and operator genes, for instance. But is not a triplet all by itself a stretch of DNA that has a well-defined function? After all, it specifies a distinct amino acid! This question is intended to serve as an illustration of some of the language difficulties that scientists have. At every stage of the exploration of the nature of the gene it has been possible to produce only an *ad hoc* definition. However, this changing concept of the gene has acted as a spur to new and fruitful lines of research. The lack of an exact definition becomes vexing only when one tries to apply the rigour of logic.

Bibliography and references

1 General reading

Cove, D. J. (1971) *Genetics*, Cambridge University Press.

Fincham, J. R. S. (1976) *Microbial and Molecular Genetics*, English Universities Press.

Support reading

Goodenough, U. and Levine, R. P. (1974) *Genetics*, Holt, Rinehard & Winston. (Chapters 5 (up to p. 209) 7 and 15 (up to p. 676) are particularly helpful.)

Mantis, T. and Ptashne, M. (1976) 'A DNA operator-repressor system', *Scientific American*, **234**, pp. 64–76. (An up-to-date review of what is known about the interactions between the repressor protein and the operator gene.)

Stent, G. S. (1971) *Molecular Genetics: An Introductory Narrative*, Freeman. (Note particularly chapters 5, 13, 16, 17, 18 and 20.)

Genetic Engineering

Ashby Committee (1975) Report of the working party on the experimental manipulation of genetic composition of microorganisms, HMSO.

Chedd, G. (1976) 'The making of a gene'. *New Scientist*, **61**, pp. 680–2. (A simple account of the work of Khorana's group in the chemical synthesis of a gene.)

Cohen, S. N. (1975) 'Manipulation of genes', *Scientific American*, **233**, pp. 25–33. (A comprehensible account of the history of recombinant DNA.)

Lewin, R. (1976) 'Genetic engineers ready for stage two,' *New Scientist*, **72**, pp. 86–7. (A comprehensible first primer to genetic engineering.)

2 Publications cited in the text

Avery, O. T., McLeod, C. M. and McCarty, M. (1944) 'Studies on the chemical nature of the substance inducing transformation of pneumococcal types. I. Induction of transformation by a deoxyribonucleic acid fraction isolated from *Pneumococus* type III', *J. Exp. Med.*, **79**, pp. 137–58.

Beadle, G. W. and Tatum, E. L. (1941) 'Genetic control of biochemical reactions in *Neurospora*', *Proc. Nat. Acad. Sci.*, **27**, pp. 449–506.

Benzer, S. (1957) The elementary units of heredity, in *The Chemical Basis of Heredity*, W. D. McElroy and B. Glare (eds), John Hopkins Press, pp. 70–93.

Britten, R. J. and Davidson, E. H. (1969) 'Gene regulation in higher cells: a theory', *Science*, **165**, pp. 349–57.

Schrödinger, E. (1945) *What is Life?*, Cambridge University Press, New York. (A thought-provoking and extremely interesting book by a physicist intrigued by the phenomenon of life.)

3 References to S100

1 Units 15 and 16 *Cell Dynamics and the Control of Cellular Activity*

2 Unit 10 *Covalent Compounds*

3 Unit 17, *The Genetic code: Growth and Replication*

4 Unit 13, *Giant Molecules*

Acknowledgements

Grateful acknowledgement is made to the following for figures used in this Unit:

Figures 6, 7 and 8 from S. Benzer (1961) *Proceedings of the National Academy of Sciences*, **47**, 410; *Figure 16* redrawn from Lewin (1974) *Gene Expression I*, Wiley; *Figure 17* redrawn from E. Terzaghi *et al.* (1966), *Proceedings of the National Academy of Sciences*, **56**, 500–7; *Figure 20* from A. S. Sarabhai *et al.* (1964) 'Colinearity of the gene with polypeptide chain', *Nature*, **201**, 13; *Figure 21* from A. Gaven (1968) 'Sense and nonsense in the genetic code' from *Science*, **160**, 152. Copyright 1968 American Association for the Advancement of Science.

7 Cytoplasmic Inheritance

Contents

List of scientific terms used in Unit 7

Introduced in S100*	Developed in this Unit	Page No.
bacteria	absorption spectra	292
chloroplast	acriflavine dye	291
chromosomes	aerobic	291
clone	anaerobic	291
density gradient centrifugation	autogamy	323
diploid	biogenesis of chloroplasts and mitochondria	310, 316
DNA		
DNA polymerase	chloroplast DNA (CDNA)	305
dominant	cytochromes	292
genotype	cytohet	300
haploid	cytoplasmic inheritance (also called extra-chromosomal or non-nuclear inheritance or non-Mendelian patterns of inheritance)	290–328
heterozygous		
homozygous		
meiosis	dextral–sinistral coiling in *Limnea*	326
mitochondrion	endosymbiosis	290
mitosis	iojap inheritance	308
mRNA	mitochondrial DNA (MDNA)	305
mutation	neutral petite	295
phenotype	petite mutant	291
protein synthesis	poky mutant	296
recessive	pollen sterility	327
RNA	polygenic	293
RNA polymerase	respiratory chain	292
transcription	respiratory-deficient mutant	292
translation	streptomycin	298
tRNA	suppressive petite	295
virus		

* The Open University (1972) S100 *Science: A Foundation Course*, The Open University Press.

Objectives for Unit 7

After studying this Unit, you should be able to:

1 Define, recognize the best definition of, and place in the correct context, the items in the list of scientific terms above.

2 Construct genetic hypotheses to explain segregation patterns that cannot be accounted for by nuclear gene segregation.
(ITQs 3, 4 and 5; SAQs 1, 2, 3, 4 and 5)

3 Devise experiments to test the hypotheses.
(ITQs 1 and 3; SAQs 1, 2, 3, 4 and 5)

4 List and contrast the biological and technical features that permit the analysis of cytoplasmic systems in *Saccharomyces cerevisiae*, *Neurospora crassa*, *Paramecium*, *Chlamydomonas* and *Zea mays*.
(SAQs 1, 2, 3, 4, 5 and 6)

5 Present evidence to support the assignment of some genes to mitochondrial or chloroplast DNA.
(ITQ 2; SAQs 2, 3 and 6)

6 List the similarities and differences among the genetic systems of mitochondria, chloroplasts and prokaryotes.
(SAQ 6)

7 Evaluate the evidence that suggests that eukaryotic organelles evolved from ancestral prokaryotic forms.
(SAQ 6)

8 Explain why the inheritance of certain cytoplasmic features in *Paramecium* and the inheritance of sensitivity to CO_2 in *Drosophila* are not regarded as examples of 'true' cytoplasmic inheritance.
(ITQs 4 and 5; SAQ 4)

9 Explain the inheritance of dextral and sinistral shell-coiling in the snail *Limnea*.
(ITQ 6)

10 Explain the significance of cytoplasmically inherited pollen sterility in the production of hybrid seeds in agricultural crops.
(SAQ 5)

Study guide for Unit 7

In this Unit we shall try to show you how the basic tools and principles of genetic analysis, together with present-day biochemical techniques, can be used to advantage in solving some intriguing, and more unusual, genetic problems—those concerned with cytoplasmic inheritance.

When you read the text, concentrate particularly on the key questions posed at the beginning of each Section and then see how the application of genetic analysis and/or biochemical techniques can help you to resolve whether a given phenotypic character is determined by a chromosomal gene or an extra-chromosomal gene. For example, unusual segregation patterns (i.e. those that are non-Mendelian in their behaviour) often give you a clue.

Probably most important is Section 7.1.3 which deals with cytoplasmic inheritance in *Chlamydomonas*, because the experiments reported here go further than most in demonstrating not only the existence of cytoplasmic genes, but also that it is possible to map such genes from recombination analyses. This Section is also important because the development and rationale for the research work with *Chlamydomonas* are discussed by Professor Ruth Sager in Radio programme 7.

As the Unit gives prominence to cytoplasmic inheritance in which chloroplasts and mitochondria are known to be involved, there is perhaps rather more biochemistry than in many of the other Units in this Course. Do not get too bothered if you cannot remember all the biochemical details; this is not essential to achieve the Objectives. However, we include many of the biochemical details to provide the necessary framework for some of the experiments outlined in the Unit. What is important is that you understand the general points being made and gain a perspective on the various aspects and implications of cytoplasmic inheritance. For this reason, we have tried to avoid an encyclopaedic approach in which obscure phenomena are simply listed and described.

Recently, there has been renewed interest in and speculation about the origin of mitochondria and chloroplasts in eukaryotic cells. We discuss some of the current theories in Section 7.5.2. However, if you are running short of time, this is a Section that may be omitted, although we hope you will read it and find it of interest.

Finally, the ITQs and SAQs are included to help you to judge how well you have understood the material in this Unit and, therefore, how well you have met its Objectives. We urge you to try the ITQs as you come to them and to complete the SAQs at the end of the Unit.

7.0 Introduction to Unit 7

You have already seen, from the previous six Units, that the existence of genes as segments of nucleic acid molecules, usually located in chromosomes and controlling phenotypes in a known and predictable fashion, is amply demonstrated on sound, observable and verifiable criteria. However, the establishment of such a chromosomal mechanism of inheritance does not necessarily preclude a role for extra-chromosomal or non-nuclear components in the cytoplasm. Ever since the re-discovery of Mendel's work by Correns and others in 1900 (see *HIST**, Section H.3.1) the question of whether all genes reside in the nucleus has been an intriguing one. Although much evidence for the existence of extra-chromosomal or cytoplasmic inheritance had been documented in the first half of this century, on the basis of genetic analysis, a better understanding of the biochemical mechanisms involved in cytoplasmic inheritance has come about more recently through advances in the techniques and knowledge of molecular biology. As a result, some (but by no means all) of the earlier ambiguities raised by genetic analysis alone have been resolved.

You may have already realized that cytoplasmic inheritance is by definition a concept that is relevant only to eukaryotes. Indeed, we can go further and say that most of the best documented and best understood accounts of cytoplasmic inheritance have been concerned with plants—particularly fungi such as yeast, *Neurospora* and *Aspergillus*, the single-celled green alga, *Chlamydomonas*, and certain cereal crops such as *Zea mays*.

In this Unit, we shall be drawing upon evidence primarily from these organisms, and concentrating in particular on the mitochondria of the non-green fungi and the chloroplasts of the green plants. (Do not forget however, that green plants also possess mitochondria.) We have made this decision because currently the biogenesis of chloroplasts and mitochondria is a very active area of research and, therefore, much more is known about these systems than many others. Also, because both organelles play such vital roles in processes for capturing and transducing energy, the information gained from their study may not only provide us with greater knowledge about their genetics as such, but also give greater insight into how the whole cell is assembled and functions.

As you work through the Unit, however, you will realize that there are still many unsolved problems, not the least of which is the major difficulty of deciding whether some of the 'abnormal' (non-Mendelian) genetic phenomena described can be attributed to cytoplasmic inheritance or not. In some instances, it is perfectly clear that the 'abnormal' effect is caused by an invading virus-like or bacterium-like particle and, therefore, is not intrinsically and permanently (i.e. from generation to generation) cytoplasmic in origin. In others, it is less clear; so much so, that we are left wondering whether structures such as chloroplasts and mitochondria, which we normally regard as permanent features of cells, might not themselves have been invading *endosymbionts* from many millions of years ago. An endosymbiont is an organism that invades and then comes to live within the cell(s) or body of another organism, from which relationship both host and invader derive mutual benefit.

endosymbiont

7.1 Non-Mendelian inheritance in fungi and *Chlamydomonas*

When looking at specific examples, we need to bear in mind two questions:

1   How can cytoplasmically inherited genotypes be detected and distinguished from those determined by the nuclear genes?

2   How can we be sure that the apparent cytoplasmic pattern of inheritance is not related to chromosome segregation, polygenic inheritance, or any chromosomal linkage group?

---

* The Open University (1976) S299 HIST *The History and Social Relations of Genetics*, The Open University Press. This text is to be studied in parallel with the Units of the Course. We refer to it by its code, *HIST*.

7.1.1   Respiratory-deficient mutants in *Saccharomyces cerevisiae*

In attempting to answer these questions, let us begin our investigation by looking at certain patterns of inheritance in baker's yeast (*Saccharomyces cerevisiae*). You will recall that the essential features of the life cycle are:

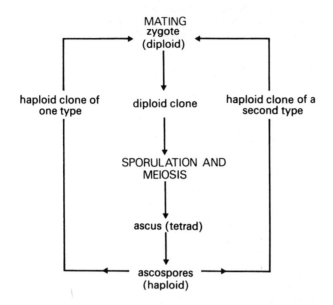

Figure 1   Essential features of the life cycle of baker's yeast.

Now, imagine that we have set up a cross between two haploid clones of different mating types *a* and α. *a* and α are a pair of chromosomal alleles that control the mating system, permitting individual cells of both types to fuse and form a diploid zygote. Eventually, the diploid clone produced from this zygote will undergo meiosis.

QUESTION   How will the alleles *a* and α segregate in the tetrad (ascus) produced?

ANSWER   In a ratio of 2*a*:2α (although the 4 meiotic products may not be in any predictable order).

Similarly, if the experiment is repeated using another nuclear marker such as the auxotrophic mutant *arg* (arginine-requiring), the meiotic products always reveal a consistent segregation ratio of 2 *arg*$^+$ (wild type): 2 *arg*.

When, in 1949, Ephrussi and his colleagues in Paris discovered a '*petite*' mutation in yeast that did not conform to this segregation pattern, it aroused considerable interest.

petite mutant

Before outlining Ephrussi's experiment, we should explain that yeast cells are very adaptable in their requirements for growth. Under *anaerobic* conditions (in the absence of oxygen), they are able to derive their energy for growth and cell division by fermenting sugars, particularly glucose. Under *aerobic* conditions (with oxygen present), however, they can use not only glucose as an energy source, but also organic molecules such as glycerol, lactate and ethanol, which contain fewer carbon atoms than glucose. These non-fermentable organic molecules can provide energy under aerobic conditions, because they can be oxidized via the respiratory chain within the mitochondria.

Figure 2(a)   Wild-type and petite colonies growing on glucose agar and stained with tetrazolium, which is taken up only by the wild-type colonies; these appear larger and darker (actual size).

The mutation discovered by Ephrussi and his colleagues manifested itself in the inability of the mutant cells to grow on these non-fermentable carbon sources. Indeed, even on glucose, the mutant cells grow more slowly and as a result form small or petite colonies (see Fig. 2(a)). The frequency of the spontaneous petite mutation is in the order of 0.1 per cent. The frequency, however, can be increased by exposure of wild-type yeast colonies to ultraviolet light or to the mutagenic dye, *acriflavine*. Following treatment with acriflavine at a concentration of 1 part in 300 000, virtually all the normal cells give rise to petite cells. Haploid and diploid cells show the same degree of susceptibility to the mutagenic source. The step-by-step identification of petite colonies is shown in Figure 2(b) (*overleaf*).

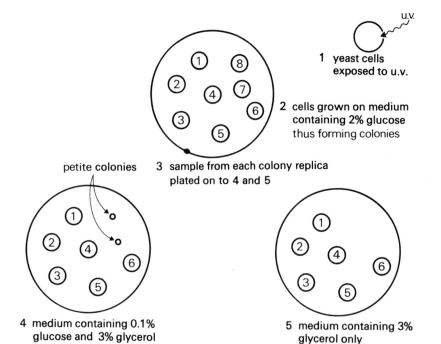

petite colonies

4 medium containing 0.1%
glucose and 3% glycerol

5 medium containing 3%
glycerol only

Figure 2(b) The identification of petite colonies.

QUESTION Why did colonies 7 and 8 grow on plate 4 but not on plate 5?

ANSWER Because the selective medium contained only a non-fermentable carbon source, whereas the small percentage of glucose in the medium of plate 4 allowed some growth of petite colonies.

One hypothesis that can be advanced to explain why petite colonies are unable to grow on an unfermentable carbon source such as glycerol even under aerobic conditions is that either the mutant strain lacks mitochondria, or it is 'deficient' in some of its mitochondrial components.

One way of testing this hypothesis is to examine the *absorption spectra* of the respiratory pigments, the *cytochrome* proteins, in both wild-type and petite cells. Because cytochromes c, b, and a + a₃ are important components of the respiratory chain (housed in mitochondria) of all organisms, the absorption spectra might provide us with a clue to the problem.

absorption spectra
cytochromes

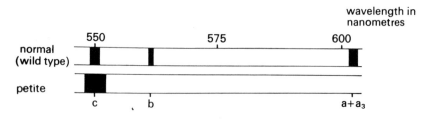

Figure 3 Absorption spectra for the reduced cytochromes obtained from normal (wild-type) and petite yeast cells

Figure 3 shows the absorption bands for the cytochromes obtained from normal (wild-type) and petite cells, respectively. (Note that each cytochrome pigment absorbs light from one wavelength only.)

Data from absorption spectra show the complete absence of cytochromes b and a + a₃ in petite cells, although cytochrome c is still strongly represented. It seems likely that petite is a *respiratory-deficient mutant*, which induces changes in the *respiratory chain* (the electron transport chain) within mitochondria.

respiratory-deficient mutant
respiratory chain

In their early experiments using acriflavine to induce the petite phenotype, Ephrussi and his colleagues noticed that acriflavine was highly specific in its mutagenic action on the hereditary determinant of petite and, as we have already noted, the effect of the dye was the same with both haploid and diploid yeast cells. However, it apparently had no effect on other hereditary factors under the experimental conditions employed.

QUESTION Why does this finding suggest that the hereditary determinant of petite may not be a nuclear gene?

ANSWER The situation is unusual as nuclear genes characteristically mutate at random when exposed to mutagenic agents. Also, most mutagens have a more adverse effect on haploid than on diploid cells (here the petite mutagenic effect applies equally to haploid and diploid cells) and the frequency of mutagenesis was atypical compared with other mutagenic agents.

To investigate the petite phenomenon further, Ephrussi carried out a genetic analysis. First, haploid wild-type yeasts of mating type a were crossed with haploid petites of mating type α. The resulting ascospores produced in each tetrad showed a segregation ratio of 2a:2α, but *all* the ascospores gave rise to normal colonies. In other words, the segregation ratio of the wild type to the petite was 4:0, and the petite phenotype had apparently vanished (see Fig. 4).

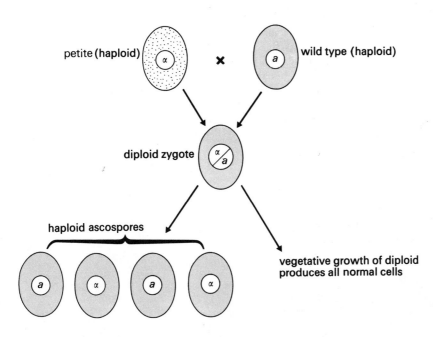

Figure 4 A cross between the wild type and a neutral petite.

The petite character, therefore, behaves in a non-Mendelian way.

QUESTION Why is it not possible to attribute the petite character to a recessive gene? (*Hint*: Look at the life cycle of yeast; Fig. 1 on p. 291 and *Life Cycles**.)

ANSWER Because the ascospores are haploid.

To demonstrate that the wild-type phenotype is retained from one generation to the next, wild-type cells of the F_1 generation were chosen at random and backcrossed to the original petite parent; all the progeny were normal. Furthermore, Ephrussi and his colleagues carried out four further backcrosses using three pairs of unlinked nuclear genes ('markers'). In addition to the mating types a and α, two unlinked auxotrophic mutants were used, one arginine-requiring and the other thiamine-requiring. As with the mating types a and α, these auxotrophic mutants each gave 2:2 segregation ratios, with very rare exceptions. In contrast, the segregation ratio for the wild type to the petite was nearly always 4:0. In only 5 ascospores out of 596 investigated did the petite phenotype appear. Ephrussi suggested that because the spontaneous mutation rate from the wild type to the petite was high, the five exceptional ascospores were new petite mutants. Although we have shown that the petite phenotype segregates in a non-Mendelian way, we have not *proved* at this stage that the petite mutation is non-nuclear in character.

For example, it is just possible that the 5 exceptional petite ascospores were segregants of a *polygenic* system. This possibility was investigated and shown to be extremely unlikely by one of Ephrussi's colleagues, P. L'Heritier, on the basis of a complex statistical analysis. Details of the analysis need not concern us in this Course, but L'Heritier showed that the incidence of 5 petites out of 596 progeny

polygenic

* The Open University (1976) S299 LC *Life Cycles*, The Open University Press. This folder, containing details of organisms mentioned in the Course, is part of the supplementary material for the Course.

would require more than 20 independently segregating nuclear genes to give the rare segregation observed. Also, he argued further that the probability that acriflavine would induce mutation in more than 20 unlinked nuclear genes was exceedingly remote, as Ephrussi and his colleagues had observed no other mutation induced by acriflavine.

Further confirmation that the petite character was not nuclear in origin came from studies by Wright and Lederberg. These workers used a strain of yeast whose mating system allows a relatively high degree of cytoplasmic 'mixing' and vegetative budding *before* the fusion of the nuclei from the two mating types. The cytoplasmic 'mixing' between cells and vegetative budding are shown diagrammatically in Figure 5. You will notice that the two mating colonies also possess auxotrophic markers that are known to be determined by nuclear genes.

During the interval before the fusion of nuclei, some of these buds contained either one or other type of parental nucleus. By careful dissection, therefore, it is possible to obtain a few cells that receive cytoplasm from both parents, but a nucleus from only one. When these cells are allowed to grow vegetatively and to form clones, some cells develop only the wild-type phenotype, whereas others develop only the petite phenotype, irrespective of the nuclear marker obtained from one or other parent. Peculiarly, too, in this yeast strain, the budding (haploid cells) show fusion with each other, so that it is possible to generate both heterozygous diploid clones (see Fig. 5,B) and homozygous diploid clones (i.e. A and A′ in Fig. 5).

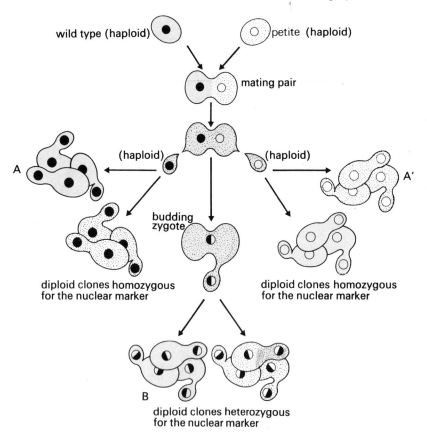

Figure 5 A strain of yeast whose mating system allows a relatively high degree of cytoplasmic 'mixing' and vegetative budding *before* the fusion of nuclei. The original cross is made between two haploid forms of yeast cell. These cells differ in two different auxotrophic genes, known to be located in the nucleus; the haploid cells also differ in their cytoplasm, one showing the petite character and the other the normal wild-type character. Haploid buds can be dissected off early from the 'mating', after cytoplasmic mixing but before nuclear fusion. These buds form homozygous diploid clones as a result of the fusion of identical haploid nuclei. The cytoplasmic character, however, either reveals itself in clones that are wholly wild-type or clones that are wholly petite, in approximately equal numbers.

If buds are not dissected early, both cytoplasmic mixing and nuclear fusion occur. Clones from this line are also diploid, but heterozygous with respect to the nuclear gene markers. The cytoplasmic character, however, once again reveals itself in colonies that are either wholly petite or wholly wild-type.

QUESTION How do the results of this experiment help to confirm Ephrussi's earlier conclusions that the petite phenotype is subject to extra-nuclear genetic transmission?

ANSWER The experiment shows that the petite character behaves quite differently from the auxotrophic (nuclear) gene markers used (i.e. it is non-Mendelian in character).

The experiment was repeated with other nuclear gene markers and each time the petite character showed a behaviour pattern that was not compatible with a Mendelian pattern of inheritance. At this point, therefore, we can say that Ephrussi's early experiments and those of Wright and Lederberg have helped us to answer the first question raised in Section 7.1, and L'Heritier's statistical analysis has to a considerable degree helped to answer the second.

So far, we have given the impression that there is only one type of petite. In 1955, Ephrussi and his colleagues found a second class of cytoplasmic petites, which they called *suppressive* (or dominant) *petites*, to distinguish them from the first class of *neutral* (or recessive) *petites**.

suppressive petite
neutral petite

QUESTION If the suppressive petite is crossed with the wild type, what will be the expected ratio of the mating types *a* and *α* in the ascospore tetrad if a Mendelian pattern of inheritance is operating?

ANSWER $2a:2\alpha$ (i.e. the predicted Mendelian segregation for nuclear genes).

In Ephrussi's experiments the nuclear genes once again behaved according to a Mendelian pattern of inheritance, but the suppressive petite phenotype yielded a ratio of almost 0 wild type:4 petite (i.e. a non-Mendelian segregation). This situation is summarized in Figure 6.

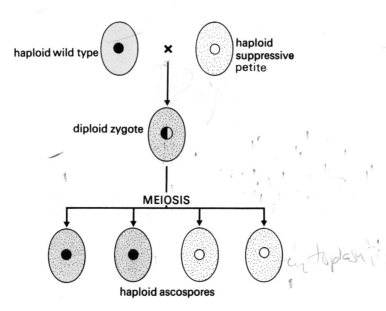

Figure 6 A cross between the wild type and a suppressive petite. ● denotes nuclear gene *a* and ○ denotes nuclear gene *α*; dotted cells represent suppressive petites (dominant) and shaded cells represent the wild type.

We stressed a ratio that was *almost* 0 wild type:4 petite because Ephrussi found that suppressiveness is never 100 per cent. At least 1 per cent of the resulting diploid cells on fusion are normal and able to produce ascospores that are all normal also. The diploid suppressive petites, on the other hand, are unable to sporulate, although they are capable of vegetative growth by budding.

QUESTION What function, present in normal diploid cells, but absent in diploid petite cells, appears to be essential for sporulation? (*Hint*: Recall the comparison of data from the absorption spectra of normal and petite cells.)

ANSWER Aerobic respiration via fully functional mitochondria seems to be essential for sporulation.

---

* We shall hereafter refer to the two types of petites as 'suppressive' and 'neutral', because we do not wish to imply competition between allelic units. The 'dominance' or 'recessiveness' is at the level of transmission in zygotes.

Further investigations have now revealed that the degree of suppressiveness can range from 0–99 per cent, depending on the strain of yeast being used. It seems that in addition to the petite character, the degree of suppressiveness itself is inherited cytoplasmically. For example, a population of yeast cells that is, say, 50 per cent suppressive, will transmit this condition to its clonal descendants, whereas another that is 80 per cent suppressive will likewise transmit this condition to its clonal descendants. An explanation for this phenomenon is difficult, but you might bear in mind that at cell division the nuclear material is distributed precisely between daughter cells, whereas the distribution of mitochondria is less precise.

For the moment, let us leave neutral and suppressive petites in yeast to see whether respiratory-deficient mutants akin to petites occur in other organisms.

7.1.2 Respiratory-deficient mutants in *Neurospora crassa*

A slow-growing form of the red bread-mould, *Neurospora crassa*, called '*poky*' is suspected of being a respiratory-deficient mutant.

poky mutant

> QUESTION What kind of experimental evidence (already discussed) would support this suspicion?

> ANSWER Evidence from absorption spectra of the respiratory chain cytochromes as shown in Figure 3 on p. 292.

When an analysis of this kind is carried out on the wild type and then on the poky form of *Neurospora*, the poky form, like the petite mutant in yeast, is found to be deficient in the mitochondrial cytochromes b and a $+$ a$_3$ although the concentration of cytochrome c is some 15 times higher in the poky form.

Having obtained some evidence that the poky phenotype is the result of a respiratory-deficient mutant in *Neurospora*, we now need to test whether the mutation is nuclear or cytoplasmic in origin. First, look at the life cycle of *Neurospora* (see *Life Cycles*).

The main points to notice are the fusion or fertilization of a haploid conidial cell ('male') of one mating type with a haploid trichogyne cell ('female') in the proto-perithecium of the opposite mating type and the later meiotic division followed by a single mitosis, giving rise to eight ascospores.

> QUESTION If the following reciprocal crosses are made

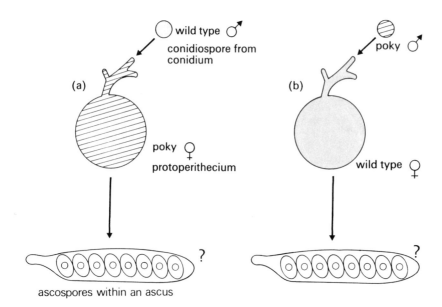

ascospores within an ascus

Figure 7

what will the segregation ratio be between the wild type and the poky form if nuclear inheritance is involved?

> ANSWER If poky is determined by a nuclear gene, then the ascospores will segregate in a ratio of 4 wild-type:4 poky.

296

In fact, when these reciprocal crosses are carried out, in (a) all the ascospores are phenotypically poky, whereas in (b) all the ascospores are wild-type.

> QUESTION What appears to be the major factor in determining the type of ascospore?

> ANSWER The 'maternal' cell (trichogyne cell in the protoperithecium) rather than the 'paternal' cell (conidium).

> QUESTION Have we now demonstrated unequivocally that poky is cytoplasmically inherited?

> ANSWER No. We have not actually seen how alleles for a known nuclear gene ('marker') behave in conjunction with poky.

As it happens, we can be fairly sure that poky is inherited uniparently via the cytoplasm, because when the reciprocal cross is carried out with a number of nuclear gene markers, alleles of the gene markers segregate in a ratio of 4:4, whereas the cytoplasmic phenotype of these ascospores is either poky or wild-type according to the maternal phenotype selected (see Fig. 8).

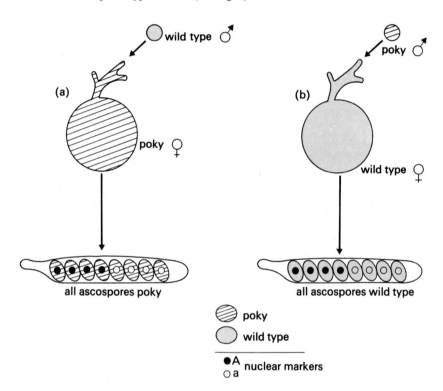

Figure 8   The results of reciprocal crosses between (a) poky (protoperithecium) and wild type (conidiospore) and (b) wild type (protoperithecium) and poky (conidiospore). The nuclear gene markers segregate in the predicted ratio of $4A : 4a$; the cytoplasmic character produces in (a) ascospores that are all poky or in (b), all wild type.

The examples that we have selected so far have shown you how genetic analysis reveals the probable occurrence of cytoplasmic inheritance, as the loss of cytoplasmic function observed does not conform to a Mendelian pattern of inheritance. Also, the indications are that the slow-growing respiratory-deficient mutants of certain fungi are caused by defects in their mitochondria. However, we have not yet demonstrated that the loss of cytoplasmic function is determined by genes residing in the cytoplasm, and we shall be returning to this problem later.

To conclude this discussion of non-Mendelian inheritance, let us examine a different system in which again certain phenotypes appear to be cytoplasmically inherited, but this time these can be attributed to defects or changes in chloroplasts.

7.1.3   Cytoplasmic inheritance in *Chlamydomonas reinhardi*

The motile, green, single-celled alga called *Chlamydomonas reinhardi*, has been subjected to extensive genetic analysis, because of the ease of investigating single-celled, yet sexually reproducing, organisms (c.f. also yeasts) with short life cycles. As you will see in this Section, this amenability has enabled research workers to identify mutants of both nuclear and cytoplasmic genes and as a result to establish a workable genetic system at both nuclear and cytoplasmic levels in which recombination and mapping can be carried out. One result has been the discovery of the first cytoplasmic linkage group.

Let us begin with a brief resumé of the life cycle of *Chlamydomonas* (Fig. 9).

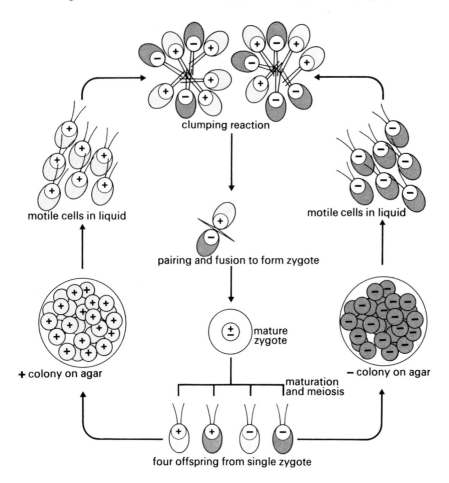

Figure 9 A summary of the life cycle of *Chlamydomonas reinhardi* showing a cross between two mating types *mt*⁺ and *mt*⁻ controlled by a single nuclear gene and another independently assorting nuclear gene.

You will see from Figure 9 that two mating types mt^+ and mt^- are involved, mating being controlled by a single nuclear gene. Also each mating type carries another unlinked nuclear gene, whose alleles are denoted by light and dark shading. Notice particularly that after isogamous* fusion of the two mating types, the four haploid offspring from the meiotic division of the diploid zygote show independent assortment in a 1:1:1:1 ratio, as predicted by Mendel's laws.

A major advance in *Chlamydomonas* genetics was the discovery that the antibiotic *streptomycin* is mutagenic for many cytoplasmic genes, but not detectably for nuclear ones (c.f. acriflavine and yeasts). The deleterious effect of streptomycin had first been noticed by von Euler in 1949, when he watered plants with a streptomycin solution and found that they developed colourless leaves as a result of irreversible damage to the developing chloroplasts.

streptomycin

In 1954, Sager was able to isolate a mutant strain of *Chlamydomonas* that had a high level of resistance to streptomycin when cultured on streptomycin (500 μg/cm³) and agar plates. She called this strain *sr*†, this being an abbreviated form of *streptomycin-resistant*. The counterpart to *sr* was referred to as *streptomycin-sensitive* (*ss*) because of its inability to grow on streptomycin and agar.

When the streptomycin-resistant strain (*sr*, *mt*⁺) was backcrossed with the streptomycin-sensitive (*ss*, *mt*⁻) or 'normal' strain, the results in the F₁, together with the four F₁ backcrosses, were as shown in Figure 10.

---

* Isogamous means, literally, gametes of equal size, that is, both contribute equal amounts of cytoplasm as well as nuclear material.

† NB In Units 3 and 4, p. 168, we give *str*ʳ and *str*ˢ as examples of symbols for streptomycin-resistant and streptomycin-sensitive, but as Professor Sager refers to *sr* and *ss* in Radio programme 7, we have decided to follow her notation here.

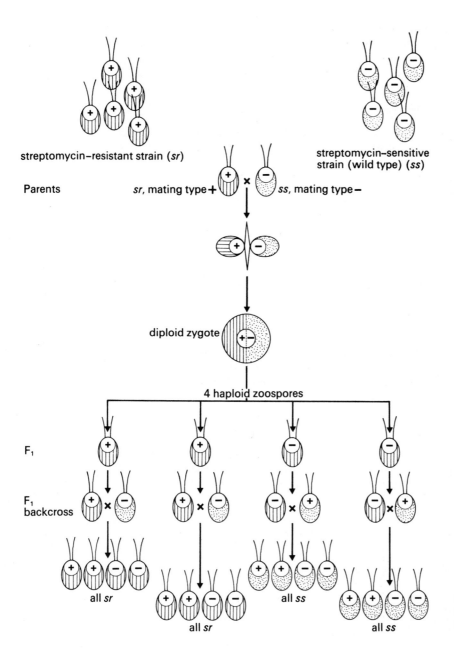

streptomycin–resistant strain (*sr*)

streptomycin–sensitive
strain (wild type) (*ss*)

Parents   *sr*, mating type **+**   *ss*, mating type **–**

diploid zygote

4 haploid zoospores

F₁

F₁
backcross

all *sr*

all *sr*

all *ss*

all *ss*

Figure 10 The inheritance of
streptomycin resistance in
Chlamydomonas.

QUESTION How does resistance or sensitivity to streptomycin seem to be
inherited?

ANSWER By uniparental inheritance from the (mt^+) mating type only.
Although the nuclear genes conferring mating types segregate in the ratio of
$2:2$, the streptomycin resistance segregates as either $4:0$ or $0:4$ depending
on the phenotype of the mt^+ mating strain.

Once again, the possibility of polygenic inheritance was eliminated by Sager and her
colleagues on the basis of four further backcrosses; in all, the uniparental pattern
of inheritance for *sr* and *ss* was confirmed and cytoplasmic inheritance appeared to be
involved. Nevertheless, to establish the existence of cytoplasmic genes, we need to
show that segregation and recombination occur between such genes. Uniparental
inheritance, however, is distinctly disadvantageous for the investigator when it
comes to applying recombination analysis and conventional mapping procedures.

ITQ 1 Can you explain why this is?

The answers to the ITQs are on p. 331.

Not only is uniparental inheritance disadvantageous from the point of view of
recombination analysis and mapping, but in *Chlamydomonas* it is a puzzle because
mating is isogamous. At least in higher plants and animals, parental inheritance
can be attributed to the fact that the contributions made by the cytoplasm of the

female and male gametes to the fertilized egg are unequal, the female contributing more (c.f. also the *Neurospora* experiments earlier, in Section 7.1.2). This cannot be so in *Chlamydomonas* where both mating types contribute equal amounts of cytoplasm because there is complete fusion of the isogametes. What mechanism, therefore, is operating to inhibit the other parental (mt^-) genotype?

At present, geneticists do not know the answer.

Nevertheless (as with many aspects of biology!), there are spontaneous exceptions to the rule of uniparental inheritance in *Chlamydomonas*. Sager found that in large numbers of crosses between streptomycin-sensitive mating type plus (+) and streptomycin-resistant mating type minus (−), 1 per cent of the zygotes transmitted the streptomycin-resistant allele (*sr*) to the progeny.

Later, Sager isolated another mutation with a non-Mendelian pattern of segregation that was streptomycin-dependent (*sd*), that is, it would not grow on agar unless streptomycin was present, so that a cross, *sd, mt^+ × ss, mt^-* gave zygotes and then further progeny that all grew well because *sd* was transmitted to them from the mt^+ parent. When this cross was repeated on a medium free of streptomycin (i.e. a minimal medium), Sager expected that no colonies would grow, because the mt^+ parent would again transmit the *sd* factor. Instead, a few exceptional zygotes (0.07 per cent) were produced that segregated into *sd* and *ss* phenotypes when grown on the two originally defined media.

The reciprocal cross, *ss, mt^+ × sd, mt^-* gave similar but opposite results to those outlined above. The two experiments, together with two control experiments, are depicted in Figure 11, which shows the segregation patterns of the exceptional zygote. The situation is also shown in Table 1, opposite.

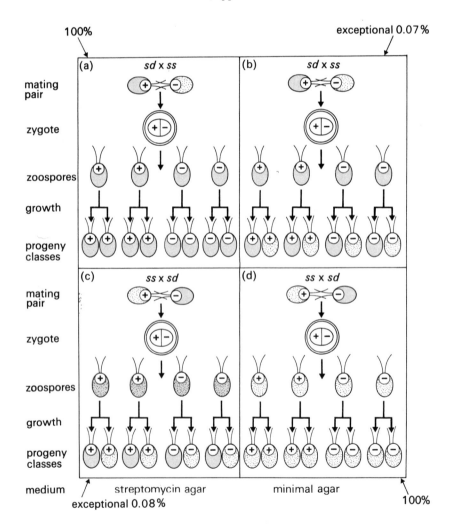

Figure 11 Reciprocal crosses between the *sd* and *ss* strains of *Chlamydomonas* showing the usual uniparental inheritance through the + mating type and the segregation patterns of the exceptional zygote. The *sd* phenotype is denoted by grey toning, the *ss* phenotype by dotting.

It seemed that in the exceptions, the zygotes, and the zoospores derived from them, possessed 'heterozygous' cytoplasm; Sager referred to these as *cytohets*. The spontaneous exceptions to the rule gave Sager and her colleagues encouragement in two directions. First, it permitted the possibility of some studies of recombinants

cytohet

300

Table 1

		% Zygote colonies formed	
	Cross	streptomycin agar	minimal agar
1	$sd, mt^+ \times ss, mt^-$	100	0.07
2	$ss, mt^+ \times sd, mt^-$	0.08	100
3	$sd, mt^+ \times sd, mt^-$	100	< 0.000 1
4	$ss, mt^+ \times ss, mt^-$	< 0.000 1	100

and, second, it showed that it might be possible to change the uniparental pattern of inheritance into a biparental pattern and thus open the way to quantitative genetic analysis of cytoplasmic genes. Eventually, Sager and Ramanis found that exposure of the mt^+ cell to ultraviolet irradiation just before mating led to biparental inheritance in about 50 per cent of cases. Consequently, quantitative recombination analysis of cytoplasmic genes was possible.

Before we examine one of Sager and Ramanis's experiments in detail, let us look at a model system based on one of their early experiments, showing an exceptional zygote of *Chlamydomonas* (Fig. 12).

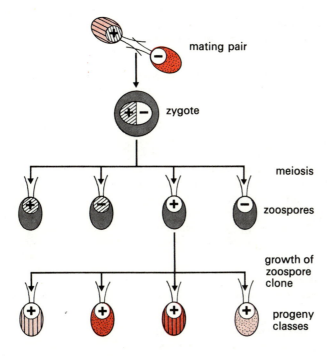

Figure 12 A model system based on the segregation pattern of an exceptional zygote of *Chlamydomonas*.

In this cross the parental cells differ from each other by (a) two nuclear genes, denoted by + and − to show the mating types and by the presence or absence of hatching in the nuclear area, and (b) two pairs of cytoplasmic genes, indicated by red tonings and *either* dotting or striping of the cytoplasm.

The mixing of the cytoplasm in the zygote and zoospores is denoted by grey toning.

In this cross, one parental cell differs from the other by two unlinked nuclear genes and also by two pairs of cytoplasmic genes. The zygote is diploid, containing *all* genes from both parents. From this mating, which would produce 16 classes of progeny, only 4 are shown from the segregation of the exceptional zygote.

QUESTION What is unusual about the segregation of cytoplasmic genes compared with the segregation of nuclear genes?

ANSWER Cytoplasmic genes do *not* segregate at meiosis. The unlinked nuclear genes, as predicted by chromosome behaviour, segregate independently in a 1:1:1:1 ratio as a result of meiosis. The cytoplasmic genes segregate during the *mitotic* divisions of each zoospore clone after meiosis, giving four classes of progeny.

The progeny classes are unusual because, in addition to the two parental types, there are two recombinant types in each zoospore clone.

QUESTION Does this result indicate that each zoospore is initially heterozygous for both pairs of cytoplasmic genes?

ANSWER Yes. Despite the fact that each zoospore is a haploid cell from the point of view of nuclear genes, the cytoplasm *initially* is 'heterozygous'.

These results based on actual experiments showed for the first time that:

(a) recombination does occur between cytoplasmic genes;

(b) segregation of cytoplasmic genes does *not* occur during meiosis, but during vegetative growth (at least in *Chlamydomonas*);

(c) cytoplasmic genes like nuclear genes have the properties of stability, mutability, maintenance of identity in heterozygotes, segregation of alleles, and allow a distinction to be made between genotype and phenotype.

To reinforce these points, let us return to Figure 12, but this time we shall substitute the cytoplasmic genes actually used by Sager and Ramanis.

The original cross involved two different acetate-requiring cytoplasmic mutants $ac1$ and ac ; the sr (streptomycin-resistant) mutant and the ss (streptomycin-sensitive) mutant, together with the nuclear mating types mt^+ and mt^-. The cross was:

$$ac1, sr, mt^+ \times ac2, ss, mt^-$$

This, when plated on agar containing streptomycin and acetate, gave rise to progeny that were all uniformly like the mt^+ parent (i.e. $ac1, sr$), thus showing the usual uniparental pattern of cytoplasmic inheritance.

The exceptional zygotes as shown in Figure 12 were picked up by plating on various selective media, so that the four classes of progeny shown in Figure 12 correspond to $ac1, sr$; $ac2, sr$; $ac1, ss$; $ac2, ss$. Further details of this experiment are discussed by Professor Sager in Radio programme 7.

We now have strong evidence from genetic analysis for the existence of cytoplasmic genes; but do cytoplasmic genes show linkage and recombination events that permit mapping analysis?

From their earlier experiments with exceptional zygotes Sager and Ramanis had qualitative evidence that suggested certain acetate-requiring mutants and streptomycin-resistant mutants were linked. The acetate-requiring mutant of *Chlamydomonas* is one that has lost the ability to grow photosynthetically, but will do so under light or dark conditions if acetate is present in the medium. As with petite mutations in yeast, we can devise selective media that will discriminate between non-acetate-requiring (ac^+) and acetate-requiring (ac) mutants. Similarly, it is possible to discriminate between streptomycin-resistant (sr) and streptomycin-sensitive (ss) types on the same basis.

With the knowledge that ultraviolet irradiation can convert uniparental to biparental inheritance in the cytoplasm of *Chlamydomonas*, and the discovery of more cytoplasmic mutants, Sager and Ramanis were able to undertake the following experiment in 1970.

It was suspected that two *different* acetate-sensitive cytoplasmic genes $ac2$ and $ac1$ and two *different* streptomycin-resistant cytoplasmic genes ($sr3$ and $sr2$) were all linked on the same cytoplasmic 'chromosome'. Their wild-type counterparts are respectively $ac2^+$ and $ac1^+$; and $sr3^+$ and $sr2^+$.

Sager and Ramanis, therefore, set up the following cross:

$$ac2^+, ac1, sr3^+, sr2^+ \times ac2, ac1^+, sr3, sr2$$

using mt^+ and mt^- as nuclear markers.

The results of two experiments, when the progeny were scored after two mitotic doublings of the zygospores, are shown in Table 2 *opposite*. The nuclear gene markers segregated in equal numbers, although this is not shown in the Table.

> QUESTION Could these results indicate that there are four unlinked cytoplasmic genes segregating independently?
>
> ANSWER No. The figures show great variation among classes of progeny. Although we have not carried out a χ^2 test here (see *STATS**, Section ST.4), statistical analysis of the data would indicate that it is highly improbable that a random segregation of unlinked genes is operating (c.f. Units 3 and 4, Tables 5 and 6 on p. 122).

* The Open University (1976) S299 STATS *Statistics for Genetics*, The Open University Press. This text is to be studied in parallel with the Units of the Course. We refer to it by its code, *STATS*.

Table 2 (simplified from Sager and Ramanis)

		Numbers of progeny	
	Progeny type	experiment 1	experiment 2
1	$ac2^+$, $ac1$, $sr3^+$, $sr2^+$	75	32
2	$ac2$, $ac1^+$, $sr3$, $sr2$	64	36
3	$ac2^+$, $ac1$, $sr3$, $sr2$	13	7
4	$ac2$, $ac1^+$, $sr3^+$, $sr2^+$	20	15
5	$ac2^+$, $ac1$, $sr3$, $sr2^+$	13	0
6	$ac2$, $ac1^+$, $sr3$, $sr2^+$	15	6
7	$ac2^+$, $ac1^+$, $sr3$, $sr2$	7	2
8	$ac2^+$, $ac1^+$, $sr3^+$, $sr2^+$	5	3
9	$ac2^+$, $ac1^+$, $sr3$, $sr2^+$	2	0
	Totals	214	101

We must, therefore, conclude that the genes are linked and that the various classes (or types) of progeny (3 to 9) are recombinants, especially as the parental classes, (i.e. types 1 and 2) are present in the greater numbers.

QUESTION Between which two genes is there (a) the highest and (b) the lowest frequency of recombination on the basis of this data?

ANSWER (a) between the $ac2$ and $sr2$ loci (classes 3, 4, 6 and 7); (b) between the $ac2$ and $ac1$ loci (classes 7, 8 and 9).

The evidence suggests linkage and, therefore, the recombination values could be used to construct a map of the cytoplasmic genes.

QUESTION Why would fairly accurate map distances be difficult to obtain with this data?

ANSWER Not all the classes of recombinants are represented; 7 classes are missing from a possible permutation of 16. It would also be unwise to base the map on comparatively small numbers and only two crosses!

After many more experiments with different cytoplasmic mutants, fairly reliable map distances between these gene loci have been obtained. Also, present-day evidence favours a circular map similar to those for bacteria, as shown in Figure 13.

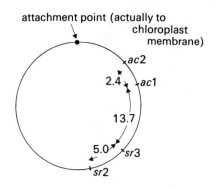

Figure 13 The proposed circular map of cytoplasmic genes in *Chlamydomonas* showing the mutant sites $ac2$, $ac1$, $sr3$ and $sr2$. Other known mutant sites are not shown here but are given in the *Broadcast Notes* for Radio programme 7.

7.1.4 Summary of Section 7.1

In this Section we have tried to provide answers to the two questions we raised at the beginning. You have seen how it is possible to detect cytoplasmically inherited traits and then to distinguish them from those determined by nuclear genes. We also presented different kinds of evidence from three different organisms, to show how we can deduce that the cytoplasmic patterns of inheritance observed are not related to chromosomal segregation, or to polygenic inheritance within the nucleus, or to any chromosomal linkage group. We have also seen some evidence from

Chlamydomonas that cytoplasmic genes do remain stable through fusion and subsequent meiosis, and that they themselves segregate during the vegetative stage after meiosis. Finally, cytoplasmic genes do show mutability, as can be seen after treatment with ultraviolet light and with dyes (e.g. acriflavine) and drugs (e.g. streptomycin). However, some of the mutagenic agents for cytoplasmic genes appear to have no detectable effect on nuclear genes.

7.2   The nature, properties and location of cytoplasmic genes

In this Section, we attempt to answer the following questions:

1   What is the chemical nature of cytoplasmic genes?

2   Where are they located?

3   What are the properties that distinguish them from nuclear genes?

7.2.1   The isolation and properties of mitochondrial DNA

In spite of the non-Mendelian segregation patterns by which cytoplasmic inheritance can be distinguished from nuclear inheritance, the evidence obtained from genetic analysis in *Chlamydomonas* indicates that the two systems have much in common. Although segregation data do not show that the genetic material is DNA, it is difficult to imagine molecules other than nucleic acids that would fulfil the role of cytoplasmic genes. Many research workers before the mid-1960s claimed that they had located DNA in the cytoplasm of a variety of cells, but, although some of this cytological evidence is convincing, conclusive proof is difficult without the isolation of the DNA. There is a major problem, however, when it comes to isolating cytoplasmic DNA, as nuclear DNA contributes between 95 and 99 per cent of the cell's total DNA, and cytoplasmic DNA only 1 to 5 per cent, depending on the type of cell. Consequently, when it was reported in 1951 that DNA was present in mitochondrial preparations isolated by differential centrifugation, the finding was generally attributed to contaminating nuclei.

Some of the more convincing cytological evidence for the existence of mitochondrial DNA came in 1963 when Nass and Nass noticed that the mitochondria of normal chick embryo cells contained fibre-like inclusions, as can be seen from one of their electron micrographs (Fig. 14).

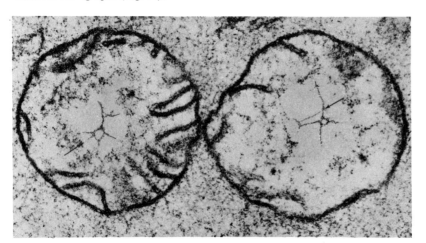

Figure 14 Intra-mitochondrial DNA fibres within chick embryo tissue fixed with osmium tetroxide ($\times$ 55 000). The arrows point to the suspected DNA fibres.

However, black specks within mitochondria on electron micrographs are scarcely proof of the existence of mitochondrial DNA!

Nass and Nass, therefore, decided to treat some of the ultra-thin sections with RNase, and others with DNase, before staining the sections with uranyl acetate.

QUESTION   What did they hope to demonstrate from these two experiments?

ANSWER   They hoped that the experiments would indicate whether DNA or RNA (or both) was/were present in mitochondria, as the enzyme RNase should degrade RNA, if present, the DNase should degrade DNA, if present.

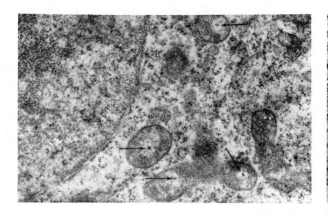 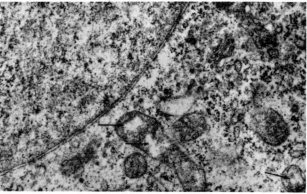

QUESTION It is to be expected that the electron micrographs taken before and after enzyme treatment would appear different. Compare Figures 15 and 16. Do they indicate that DNA is present, but not RNA?

ANSWER Yes. The fibres are probably DNA, as only DNase had any effect.

Figure 15 (*left*) The same tissue as in Figure 14, but after sectioning, treated with RNase. (× 18 000)

Figure 16 (*right*) The same tissue as in Figure 14, but after sectioning, treated with DNase. (× 18 000)

Nass and Nass then went on to show that on the same kind of evidence these DNA fibres (?) were present in the mitochondria of some 60 different organisms. Electron microscopy alone, however, does not permit a satisfactory identification of DNA.

ITQ 2 Can you suggest what further information might be obtained if uncontaminated mitochondrial DNA or chloroplast DNA can be isolated?

Isolation of *mitochondrial DNA (MDNA)*—and *chloroplast DNA (CDNA)*—has become possible for two main reasons.

mitochondrial DNA (MDNA)
chloroplast DNA (CDNA)

1 Mitochondria and chloroplasts are quite distinctive; they differ in size, shape, composition and density, not only from each other, but also from nuclei, which are larger and denser. After separation from nuclei by differential centrifugation, mitochondria (and chloroplasts) can be freed from much of the nuclear contamination by extensive washing and further centrifugation. A further refinement has been the use of special density-gradient techniques. In 1964, Schatz and his colleagues used a 'Urografin'* gradient, and found that with highly purified yeast mitochondria they could obtain a small but reproducible band within the gradient, which on biochemical assay (or staining) was found to be DNA.

QUESTION What other biochemical assay could they have carried out on this band (DNA distribution) to support the idea that it was mitochondrial in origin?

ANSWER An assay of mitochondrial enzyme activity. As Schatz and his colleagues found that the distribution of DNA in the gradient precisely paralleled the distribution of mitochondrial enzyme activity, they were fairly convinced that the DNA was associated with mitochondria.

2 It was found that DNA associated with intact mitochondria, in contrast to isolated MDNA or isolated nuclear DNA, was not degraded by added DNase.

QUESTION Can you explain this finding? How does it fit in with the earlier cytological evidence that we presented from Nass and Nass (Figs. 14, 15 and 16)?

ANSWER The DNA in intact mitochondria (unlike isolated DNA) is protected from attack by DNase by the two mitochondrial membranes. This evidence supports Nass and Nass's cytological evidence that DNA resides close to the centre of the particle.

(Note: Nass and Nass were able to remove MDNA by DNase treatment only *after* thin-sectioning of the cells, that is, the mitochondria were not intact.)

The finding that DNA within intact mitochondria was resistant to attack by DNase, although contaminating nuclear DNA adsorbed on to the outside surface of the organelle was susceptible to ready removal, provided biologists with yet another

* 'Urografin' is a commercial name for certain iodinated aromatic hydrocarbons.

means of isolating uncontaminated MDNA (and CDNA). It became possible to determine some of the general properties of MDNA, including its shape as seen by electron microscopy. The first surprising, yet distinguishing, feature was that most (perhaps all) MDNAs have a circular configuration (Fig. 17).

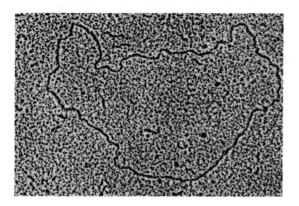

Figure 17 Isolated, circular, MDNA for a culture of mammalian cells (L-cells from mouse fibroblasts).

QUESTION In Figure 17, what is the circumference of this particular MDNA if the magnification of the electron-micrograph is × 38 000? (Give your answer to the nearest μm).

ANSWER Approximately 5.0 μm.

In fact, a large number of the eukaryotic cells investigated have MDNA with similar appearance and length. At this stage, we should draw your attention to the fact that the circularity and lengths of MDNAs bear much closer similarities to bacterial DNA than they do to chromosomal DNA. Also, these features provide us with a criterion of intactness for MDNA, and thereby allow us to estimate both the number of nucleotide bases and the relative molecular mass.

QUESTION Why is knowledge of these properties of MDNA important for the geneticist?

ANSWER It gives some indication of the coding capacity of the molecule.

Biophysical measurements show that the distance between any two base pairs is 34 nm.

QUESTION How many nucleotides will a DNA molecule of circumference (i.e. length) 5 μm contain?

ANSWER Approximately, 15 000 $\left(\text{i.e. } \dfrac{5\,000}{0.34} = 14\,700\right)$.

Physical estimates also suggest a mass of 1.92×10^6 daltons per μm of MDNA. For a circular DNA molecule of 5 μm in circumference the relative molecular mass would, therefore, be 0.96×10^7 daltons. Calculations of this kind, taken in conjunction with other evidence, indicate that each mitochondrion contains at least one circular DNA molecule, and may contain between 2 and 6.

Although we have an idea of the relative molecular mass and number of nucleotide bases present, we do not know what the coding potential of MDNA is compared with nuclear DNA.

QUESTION Would you expect the nature and sequence of bases to be the same for nuclear and mitochondrial DNA?

ANSWER Not if they are independent genetic systems carrying different information. Even closely related species often show marked differences in the relative frequencies of the bases in their nuclear DNA.

Chemical assay reveals that MDNA has a much higher adenine + thymine to guanine + cytosine base-composition ratio $(A + T):(G + C)$ than nuclear DNA. This feature can be used to separate MDNA (and CDNA) from nuclear DNA. If two or three DNAs with sufficiently different base compositions are centrifuged at very high speed for 24 hours or more in a cesium chloride density gradient, each DNA will sediment in the gradient according to its buoyant density (related to base composition). The concentration and peak absorbance (buoyant density) of each band can be obtained by ultraviolet absorbance in a spectrophotometer, provided a 'marker' of known density is also present. Figure 18 shows the gradient bands and the ultraviolet absorption spectra of nuclear, chloroplast and mitochondrial DNAs from the single-celled alga, *Euglena viridis*.

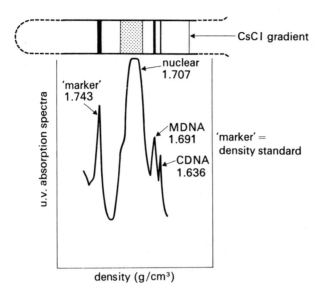

Figure 18 The gradient bands and the ultraviolet absorption spectra of nuclear, chloroplast and mitochondrial DNAs from *Euglena viridis*.

To summarize:

1 MDNA is normally a double-stranded, circular molecule with a circumference of approximately 5 μm. In yeast and other single-celled organisms it may be up to 25 μm.

2 MDNA is located near the centre of the mitochondrion; there may be more than one DNA molecule per mitochondrion.

3 It is estimated that the coding capacity of MDNA is limited to about 30 genes (15 000 nucleotides, which could code for about 5 000 amino acids representing 5×10^4 daltons of protein). Again, the coding capacity in yeasts may be greater.

4 MDNA and CDNA differ from nuclear DNA in their buoyant densities. These variations can be attributed to differences in composition. In yeast, for example, the bases A + T make up about 80 per cent of MDNA's base composition, whereas they contribute only about 60 per cent to the base composition in nuclear DNA; the remaining percentages in each case are attributed to the G + C bases.

7.3 How independent of the nucleus is the cytoplasmic genome?

There are two ways of examining this problem: one through breeding analysis; the other through molecular genetics and cell biology. In Section 7.3.1 we use the former approach and in Sections 7.3.2 and 7.3.3, the latter.

7.3.1 The inheritance of iojap in *Zea mays*

Variegation* in the leaves of some plant species is a common feature, so much so that variegated plants are often prized as horticultural ornaments. Many kinds of variegation in maize are caused by nuclear gene mutations that inhibit the normal development of chloroplasts from plastids. For example, the mutant gene 'albina' blocks the very early stages of plastid development, preventing the formation of the internal chloroplast membranes that are so essential for the production of chlorophyll. Another series of mutants called 'xantha' inhibits at a later stage in chloroplast development, blocking the synthesis of chlorophyll at different steps (see Fig. 19).

---

* Variegated leaves are leaves of higher plants that have both coloured (usually green) and colourless areas in variable amounts; the colourless areas lack chlorophyll and, therefore, do not photosynthesize.

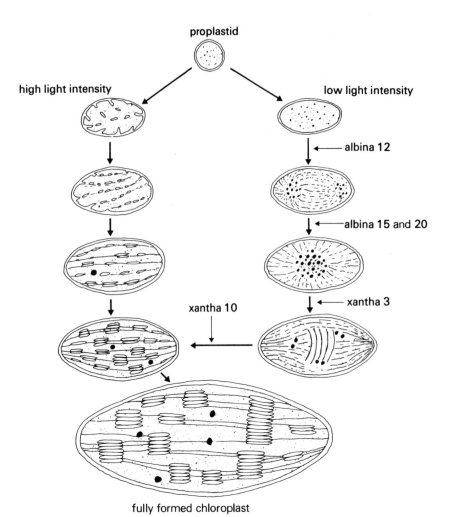

<figure>
proplastid

high light intensity — low light intensity

← albina 12

← albina 15 and 20

xantha 10 → ← xantha 3

fully formed chloroplast
</figure>

Figure 19 Chloroplast development in high and low light intensity from proplastids. We also show the points at which certain mutant genes exert their effect on chloroplast development.

iojap inheritance

In some classic studies on inheritance in maize (*Zea mays*) during the 1920s, 1930s and 1940s, first Jenkins then Rhoades used another kind of leaf variegation called '*iojap*'. Jenkins discovered 'iojap' in 1924, when he used a maize source called 'Iowa' and crossed it with a striped (variegated) variety called 'Japonica'—hence the abbreviated name 'iojap'. He went on to discover that when the normal green female was crossed with a variegated iojap (male) the F_2 progeny segregated in a ratio of 3:1 (actually, 2 498 green:782 iojap, including 12 white plants).

Maize plants that are homozygous for the recessive gene iojap (*ij*) have variegated leaves. The white areas of the leaf have cells in which the plastids are small and colourless.

Later, when Rhoades made reciprocal crosses between the variegated plants and homozygous normal plants, he obtained an unexpected result in the F_1 phenotypes. Although the cross between normal females and variegated males gave all green plants in the F_1 (as expected), the reciprocal cross between variegated females and normal males did not:

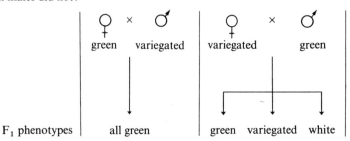

	♀ × ♂		♀ × ♂	
	green	variegated	variegated	green
F_1 phenotypes		all green	green variegated white	

ITQ 3 Why are these results unexpected in Mendelian terms?

The occurrence of the three different types of F_1 seedlings with non-Mendelian ratios made Rhoades suspect that variegation was primarily due to cytoplasmic inheritance on the maternal side, but that this was in some way directed or influenced

308

by the nuclear gene *ij* in the homozygous condition. Also, once the variegated condition had been initiated, it then continued to be inherited through the cytoplasm.

QUESTION If Rhoades's hypothesis (suspicion!) is correct, what would be the expected progeny from (a) a backcross between green F_1 plants and normal males (pollen) and (b) a backcross between variegated F_1 plants and normal males?

ANSWER

(a)

♀		♂
green		green
$\dfrac{ij^+}{ij}$	×	$\dfrac{ij^+}{ij^+}$

progeny
all green

$$\dfrac{ij^+}{ij} \qquad \dfrac{ij^+}{ij^+}$$

(b)

♀		♂
variegated		green
$\dfrac{ij^+}{ij}$	×	$\dfrac{ij^+}{ij^+}$

progeny of three phenotypic classes—green, variegated and white—in widely varying ratios, ranging from individuals wholly green to some wholly white

$$\dfrac{ij^+}{ij} \qquad \dfrac{ij^+}{ij^+}$$

QUESTION Why would it not be possible to carry out a backcross between the white phenotypes of the F_1 and the normal male (pollen)?

ANSWER White is 'sub-lethal', that is, the white seed will germinate and produce a seedling from the food reserves in the seed, but this seedling will soon die because it cannot photosynthesize; it is, therefore, incapable of producing either ovules or pollen.

It would seem, then, that the nuclear gene *ij*, when homozygous (*ij*/ /*ij*) causes variegation or striping of the leaves, but once this striped condition has been initiated, it is then inherited through the maternal cytoplasm, regardless of the genotype of the male. We can predict that the iojap gene itself is behaving according to Mendel's laws, because a closely linked marker gene segregates in the predicted ratio. Rhoades, therefore, came to the conclusion that the iojap gene was able to induce an irreversible change (mutation) in the plastids. This implies that a hereditary unit in the chloroplast has mutated under the influence of nuclear gene action. But *ij* does not induce a change in all plastids, only a certain proportion of them, which may vary from cross to cross. For example, it is possible to produce normal green plants that are homozygous for *ij* (i.e. *ij*/ /*ij*) as demonstrated below:

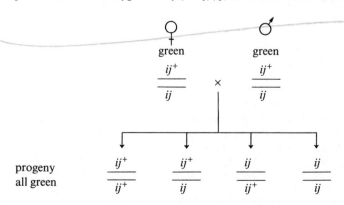

♀		♂
green		green
$\dfrac{ij^+}{ij}$	×	$\dfrac{ij^+}{ij}$

progeny
all green

$$\dfrac{ij^+}{ij^+} \qquad \dfrac{ij^+}{ij} \qquad \dfrac{ij}{ij^+} \qquad \dfrac{ij}{ij}$$

Although these experiments provide evidence of interaction between nuclear and cytoplasmic genes, they do not tell us anything about the specific metabolic step or the specific site where the mutation occurs. To try to answer our initial question at a molecular level we must turn our attention to the biogenesis of mitochondria and chloroplasts.

7.3.2 The biogenesis of mitochondria

If we are to assess the degree of autonomy exerted by cytoplasmic genes associated with certain organelles, we need first to compare the behaviour of organelles with the behaviour of the nucleus. Second, we need to examine how the organelle is made (i.e. its biogenesis).

As we have already pointed out in Section 7.2.1, if we have some measure of the potential coding information of the cytoplasmic genome, then we can begin to discover whether the organelle possesses the necessary 'machinery' to make all its structural and enzymic components.

In Section 7.1.3 we saw that segregation and recombination of cytoplasmic genes in *Chlamydomonas* occurs during the early stages of vegetative growth (mitosis) of the haploid zoospores, that is, *after* meiosis. This would suggest that the chloroplast genome has quite a degree of autonomy. However, in *Chlamydomonas*, we have an unusual situation in that we are dealing with only a *single* chloroplast—this is certainly not so in the photosynthesizing cells of higher plants.

Most cells (including *Chlamydomonas*) contain several mitochondria; many contain several hundreds. Under carefully controlled physiological conditions it seems that the number of mitochondria per cell remains constant. Nevertheless, because all cells have divided at some stage in their cycle and many cells continue to do so, some very frequently, we have the problem: 'how do new mitochondria arise in cells?'

QUESTION There are essentially two alternative explanations to this question; can you suggest what they are?

ANSWER (a) Mitochondria arise from pre-existing mitochondria (as new cells arise from pre-existing cells), that is, mitochondria are themselves capable of growth and division under the direction of their own genome.

(b) Mitochondria are assembled from 'building-block' components already present in the cell, or *de novo*, as it is often expressed, under the direction of the nucleus or the nucleus and the cytoplasmic genome.

There is, in fact, a third possibility that mitochondria are transformed or differentiate from pre-existing membranous structures, that is, by invagination of the plasma membrane or the nuclear membrane (this is a modification of (b)).

The question was answered by Luck in 1963 using the mould *N. crassa*. Choline is an important lipid constituent of biological membranes, and particularly of mitochondrial membranes. In *N. crassa*, a choline-requiring mutant is known that will grow normally, provided that choline is supplied to the fungus in the culture medium. Taking advantage of this mutation, Luck grew the mutant form of the fungus on radioactive choline (^{14}C-choline). As the radioactive label was known to be 'stable', that is, it did not get exchanged with other molecules in the cell or become a metabolic waste product and thus get eliminated, it was possible to determine the distribution of the label after successive divisions. Following an incubation period, during which the mitochondria became uniformly labelled, as determined from autoradiographic survey, Luck transferred the fungal cells to a 'cold' or unlabelled choline medium, where they were allowed to double in mass. When the cell mass had doubled, autoradiographic evidence showed that the *average* radioactive grain count in the mitochondria had diminished by *half*. Autoradiographic counting at intervals during the doubling of the cell mass revealed a *random* pattern of labelling in some 300 mitochondria. The experiment is set out in Figure 20.

310

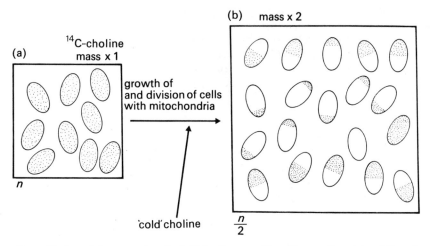

Figure 20 A model representing the initial cell mass with uniform radioactivity, followed by growth and division of the original mass to twice its size in an unlabelled medium. The mitochondrial grain counts reveal a randomized distribution in the cells, some cells showing almost 100 per cent of the original, others around 50 per cent and yet others around 25 per cent.

(a) n = grains of radioactivity determined initially.

(b) $\frac{n}{2}$ = grains of radioactivity after a doubling of cell mass, revealing the random pattern of labelling in the mitochondria.

QUESTION Why do the results of this experiment support the view that mitochondria arise by the growth and division of pre-existing mitochondria rather than *de novo*?

ANSWER The results show that mitochondria increase in mass by the continuous addition of new, choline-containing lipids to pre-existing mitochondria, followed by division in a completely *random* manner (apparently independent of nuclear division).

On a *de novo* basis we should expect that half of the mitochondria would be fully labelled and the other half unlabelled.

We now have another piece of evidence that suggests that mitochondria, like chloroplasts, exhibit a considerable degree of autonomy from the nucleus. It is difficult, however, to envisage how these organelles could have total autonomy, if cytoplasmic activities are to be co-ordinated.

Let us now take a closer look at the code and the protein-synthesizing mechanisms responsible for building the mitochondrial structure, to see whether these organelles also possess their own equally distinctive means of gene expression. Prerequisites for any protein-synthesizing system are the enzymes and components necessary to transcribe the message from DNA and translate the message (mRNA) at the ribosomes, and a source of energy in the form of ATP.

QUESTION If these prerequisites for protein synthesis are present within mitochondria as distinct entities (i.e. distinct from the nuclear–cytoplasmic system), how should we be able to recognize them? (This is a tough one, but you should be able to think of some of the answers!)

ANSWER (a) The RNAs (especially mRNA) have a structure complementary to MDNA and should be sufficiently different from analogous RNAs found elsewhere in the cell.

(b) There should be manifestations of their function in the form of specific proteins.

(c) Isolated mitochondria should be able to carry out protein synthesis.

311

Only within the last 10 years has it been possible to establish that mitochondria do contain a unique set of RNAs* and transcribing enzymes. Some of the early evidence that pointed to this fact came from work by Barnett and Brown in the mid-1960s using highly purified mitochondria from *N. crassa*. Barnett and Brown centrifuged the mitochondria in sucrose density gradients and then systematically examined some 40 or so fractions within the gradient for RNA and for cytochrome oxidase (a mitochondrial enzyme) activity. Their results are shown in Figure 21.

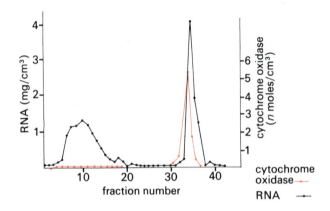

Figure 21 Cytochrome oxidase activity and RNA component associated with the purified mitochondrial fractions centrifuged in a sucrose density gradient.

QUESTION What interpretation can be put on these results?

ANSWER The mitochondria sediment out at a specific region within the sucrose gradients. This region corresponds to the area between fractions 32 and 36 in Figure 21 as deduced by cytochrome oxidase activity. There is also a peak of RNA activity within this same region. The RNA is very unlikely to be nuclear or ribosomal because the mitochondria were highly purified. The only possible RNA contamination might be from the carrying over of a small amount of tRNA located in the soluble cytoplasm. Moreover, this tRNA may be responsible for the smaller peak seen within fractions 4 to 19.

(Note: You must *not* draw the conclusion that the suspected mitochondrial RNA is in part responsible for the synthesis of cytochrome oxidase because you have no evidence at present about where cytochrome oxidase is synthesized.)

For the purpose of this Unit, we can summarize some of the evidence that suggests the existence of discrete and unique RNAs within mitochondria as follows:

1 Electron micrographs reveal the presence of ribosome-like particles within mitochondria.

2 Specific mitochondrial ribosomal particles that are only about half the size of their cytoplasmic counterparts (the sedimentation coefficients are respectively 55 S† and 80 S) have been isolated from mitochondria. Mitochondrial ribosomes also differ from their cytoplasmic counterparts in their constituent proteins; this difference is especially pronounced in multicellular organisms.

3 The 'starting signal' or chain initiating tRNA is N-formyl methionyl—tRNA in mitochondrial polypeptide synthesis (like bacteria), rather than methionyl—tRNA as in the cytoplasmic ribosomal system.

4 Mitochondrial RNAs are much more resistant to attack by RNase, both within intact mitochondria and in isolation.

5 The base composition of mitochondrial RNAs shows them to have a low percentage of G + C (only 26 per cent in yeast); this compares well with our earlier findings on base sequences in MDNA.

---

* They have tRNAs with sedimentation coefficients that differ from those of cytoplasmic tRNAs and one unusual tRNA akin to a tRNA in bacteria, namely, N-formyl methionyl tRNA.

† S = Svedberg units, used to measure the rate of sedimentation of a particle under specified conditions; the higher the value of S, the larger the particle.

6   Specific DNA–RNA hybridization studies show that mitochondrial RNAs are indeed complementary to MDNA, but when MRNAs are 'mixed' with nuclear DNA from the same cell type, no hybridization occurs.

Now, if we accept that the machinery exists within mitochondria for synthesizing protein and that it is quite separate from the machinery in other parts of the cytoplasm, we still need to know whether mitochondria do synthesize proteins, and if they do, what sort of proteins are synthesized.

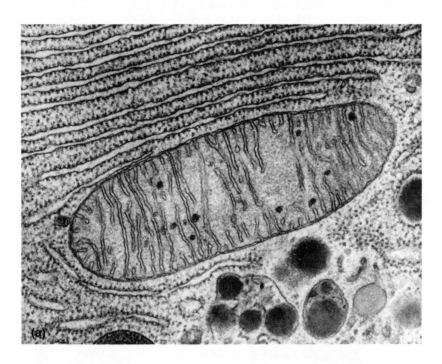

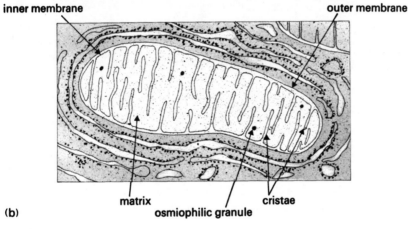

inner membrane   outer membrane

matrix   cristae

osmiophilic granule

(b)

Figure 22   (a) An electron micrograph (× 60 000) and (b) a diagram of a thin section of mitochondrion.

Figure 22 shows (a) an electron micrograph and (b) a diagram of a mitochondrion. You can see that the mitochondrion consists of an outer membrane and an inner membrane, which is intricately infolded to form cristae. Within the inner membrane is the mitochondrial matrix. The outer mitochondrial membrane is thought to play a rather 'passive' role from the point of view of transportation of solutes and solvents and metabolic activity; its main function is protection. The inner membrane, in contrast, is selectively permeable to a variety of solutes; it also houses many enzymes and the components of the respiratory chain (flavoproteins and cytochromes, etc.). The matrix contains most of the enzymes of the tricarboxylic acid cycle and fatty-acid oxidation cycle.

Having reminded you briefly about the structure and function of mitochondria, let us return to the problem of protein synthesis.

It has been known for some time that actively respiring, isolated mitochondria will take up and incorporate into their protein a variety of radioactively labelled acids from the surrounding incubation medium, although the amount of label taken up is relatively small. (Mitochondria from petite mutants of yeast will not.)

313

Early reports of this phenomenon were criticized on the grounds that the uptake was due to the presence of bacteria, as neither the mitochondrial suspensions nor the incubating media were sterile. That bacterial contamination was not involved was demonstrated by Work and his colleagues in 1967. After rearing rats under sterile conditions, they removed the late embryos by Caesarian section and isolated their liver mitochondria (also under sterile conditions). The uptake of labelled amino acids into protein took place under aerobic conditions but, in addition, they checked samples of the mitochondrial suspension on a sterile bacterial culture medium—no bacterial colonies were formed over a period of 24 hours. We must conclude, therefore, that amino-acid uptake by isolated mitochondria is a genuine phenomenon.

In experiments where isolated, but actively respiring, mitochondria were incubated with labelled amino acids, the label was found to be incorporated into some proteins, called 'structural proteins'*, of the inner mitochondrial membrane, but not into the 'soluble' enzymes of the matrix and not significantly into the proteins of the outer membrane, irrespective of the time of incubation in the label. However, when slices of kidney tissue were incubated with labelled amino acids and the rate of incorporation of label was monitored at intervals, 'structural proteins' of the inner mitochondrial membrane took up the label more intensely at first, but as time went on, the 'soluble proteins' in the matrix and the outer-membrane proteins also took up the label, until eventually the label was distributed evenly throughout the mitochondrion.

> QUESTION What implications can be tentatively drawn from the results of these two observations?

> ANSWER In the *in vivo* experiment (with tissue slices), the lag between the incorporation of label into the 'structural' proteins of the inner membrane and the incorporation of the label into other proteins in the inner membrane and parts of the mitochondrion, suggests that the proteins labelled later are synthesized extra-mitochondrially. This is supported by the *in vitro* experiment, in which only certain insoluble 'structural' proteins of the inner membrane incorporate the label, thus indicating the limited ability of the mitochondrion to synthesize protein and hence confirming the limited coding capacity of MDNA.

We have given a very brief and simple description of these experiments, but we should point out that the technical problems and interpretative difficulties presented to biologists in this field are considerable.

The picture emerging is that mitochondria are only semi-autonomous, because they depend on the nucleus for information to code for the majority of their proteins. You may recall earlier than MDNA with a circumference of 5 μm contains about 15 000 base pairs.

> QUESTION How many amino acids could be coded for and how many small proteins (of about 150 amino acids) could be produced?

> ANSWER 5 000 amino acids and, at best, about 30 small proteins.

A small enzyme like ribonuclease consists of 124 amino-acid residues, but when we consider that the components of the respiratory chain and their associated enzymes have a total relative molecular mass of several millions, it is immediately apparent

---

* 'Structural protein' is a convenient but controversial term, which really expresses our ignorance about many of the insoluble proteins that constitute the inner mitochondrial membrane. Recent evidence (1972–1974) from the mitochondria of yeast and *Neurospora* indicates that MDNA directs the synthesis of small protein sub-units, which, together with larger sub-units synthesized on cytoplasmic ribosomes, contribute to the ATPase, the cytochrome oxidase and cytochrome b complexes. For example, the cytochrome oxidase complex in *N. crassa* can be separated chromatographically into seven polypeptide sub-units of relative molecular mass: 18 000, 17 000, 13 000, 11 000, 8 000, 36 000 and 28 000. Only the first (18 000) appears to be synthesized on mitochondrial ribosomes, but we are still waiting for confirmation that this polypeptide is actually coded for by DNA.

'Soluble protein' is another term that expresses our uncertainty, this time about whether the proteins within the matrix (i.e. the tricarboxylic acid cycle and fatty-acid oxidation cycle enzymes, etc.) are actually soluble *in situ* or simply appear to be so with the techniques used to extract them from intact mitochondria.

that MDNA cannot possibly synthesize all of the mitochondrial proteins. On the other hand, some investigators have gone so far as to say that about 15 per cent of the base pairs are transcribed as ribosomal and tRNAs and, of the remaining base pairs, much is in 'meaningless' code (i.e. high A + T frequency) so that the fraction of mitochondrial proteins actually synthesized in the organelle is only about 5 per cent. Perhaps this is an underestimate, but at present we cannot be certain that the value is too low.

Another important finding is that the mitochondrial and extra-mitochondrial systems for synthesizing proteins react differently to certain antibiotics, thus allowing us to discriminate between them. For example, cycloheximide inhibits extra-mitochondrial but not mitochondrial protein synthesis of eukaryotes, yet the antibiotics chloramphenicol and erythromycin inhibit mitochondrial protein synthesis, but not extra-mitochondrial protein synthesis. In this respect the mitochondrial system shows a strong resemblance to the protein-synthesizing system of bacteria.

Taking advantage of the differences in sensitivity to these antibiotics of the protein-synthesizing systems, Linnane and his co-workers in the late 1960s carried out a series of experiments to determine the contributions made by each system to the mitochondrial proteins of yeast. Yeast mitochondria appear to contain about five times more MDNA than those of higher plants and animals. We would, therefore, expect the coding capacity of yeast to be greater as well. When bakers' yeast, *Saccharomyces cerevisiae*, was grown on a fermentable carbon source in the presence of chloramphenicol or erythromycin, the developing mitochondria* became respiratory-deficient, lacking cytochromes $a + a_3$, b and c_1. (Once again, compare this with the petite mutation where analysis reveals a deletion in the DNA.) In contrast, their content of cytochrome c and other easily extracted mitochondrial enzymes such as fumarase, acontinase and malic dehydrogenase was approximately the same, or even higher than that of the non-repressed cells. Electron micrographs of the cells grown in the presence of chloramphenicol revealed the presence of atypical mitochondria, showing apparently normal outer membranes, but only a few and poorly defined cristae (inner membranes).

QUESTION What conclusions about mitochondrial protein synthesis can be drawn from these experiments?

ANSWER The normally insoluble inner-membrane proteins such as cytochrome $a + a_3$, b and c_1 (and perhaps succinic dehydrogenase†) are synthesized by the mitochondrial system, whereas the proteins of the entire outer membrane, as well as the more easily solubilized enzymes of the matrix, are formed in the cytoplasm (and transported in some way into the mitochondrion?).

QUESTION Why was it essential for Linnane and his co-workers to grow the yeast on a fermentable carbon source?

ANSWER Energy for growth would have to be provided by a fermentation pathway because the aerobic pathway (via mitochondria) was respiratory-deficient. (Note: This parallels in many respects the situation with petites.)

Unfortunately, the results of the Linnane experiments with yeast cells do not completely agree with the results obtained from experiments looking at the incorporation of labelled amino acids into *isolated* mitochondria. Although there is agreement about the outer membrane and the dehydrogenases of the matrix, the results of experiments with isolated mitochondria do not confirm the uptake of label into the cytochromes or into succinic dehydrogenase.

The differences may be rationalized in three ways:

1 The incorporation of amino acids into isolated mitochondria might be very unrepresentative of the biosynthetic capacity of the mitochondria *in vivo*.

---

* Yeasts that are grown anaerobically have immature and poorly developed mitochondria called pro-mitochondria; when transferred to an aerobic environment they mature and develop normally.

† Succinic dehydrogenase is an important enzyme in the tricarboxylic acid cycle and electron transport system.

2   Mitochondria do not themselves synthesize cytochromes (these are made extra-mitochondrially), but chloramphenicol inhibits the synthesis of mitochondrial proteins (enzymes?) that are necessary for transporting cytochromes into mitochondria and then 'arranging' them in the inner membrane.

3   The normally insoluble cytochromes and succinic dehydrogenase *are* synthesized on mitochondrial ribosomes, but under the direction of messenger RNAs that are transcribed from the nuclear genome.

These and many other problems have yet to be resolved, but we tentatively summarize the most likely method of controlling mitochondrial assembly in Figure 23.

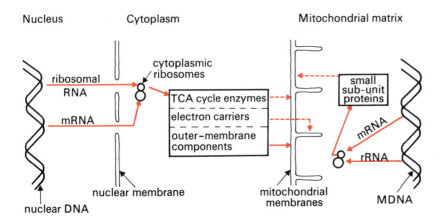

Figure 23 A suggested model for the control of mitochondrial assembly.

7.3.3   The biogenesis of chloroplasts

We conclude this Section with a brief account of the biogenesis of chloroplasts, which in many respects parallels the findings about mitochondria. For example, chloroplast DNA can be isolated and identified in much the same way as MDNA. In *Chlamydomonas*, for example, CDNA can be isolated from whole cells or from isolated chloroplasts by centrifugation in a cesium chloride gradient (Fig. 24).

biogenesis of chloroplasts

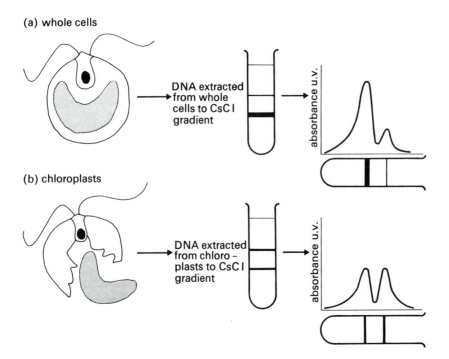

Figure 24 The isolation of CDNA from (a) whole cells and (b) isolated chloroplasts of *Chlamydomonas*.

QUESTION How would you explain the differences between the ultraviolet light absorbance peaks in Figure 24(a) and Figure 24(b)?

ANSWER The major peak in Figure 24(a) is nuclear DNA and the minor peak CDNA. When chloroplasts are isolated from nuclei as in (b), the major peak is reduced (although there is still some nuclear contamination), and the small peak is enriched several-fold.

Electron microscopic and chemical evidence suggests that CDNA like MDNA is normally circular, but longer than MDNA. In *Euglena*, for example, the circles of DNA average about 43 μm, corresponding to about 5×10^{-15} g of DNA per chloroplast. The dimensions of *Chlamydomonas* CDNA appear to be very similar, but in chloroplasts of the green unicellular alga, *Acetabularia*, lengths of CDNA up to 400 μm have apparently been found. These, of course, begin to approach the size of bacterial DNAs.

At present, far less is known about CDNA than MDNA, but CDNA like MDNA has a low percentage of (G + C) in its base composition. Nevertheless, because of its larger size, CDNA is thought to be able to code for about 3 000 different small proteins. Again, all the evidence suggests that the enzymes for DNA replication and transcription, and all of the components needed for translation of mRNA (e.g. ribosomes, tRNAs, enzymes, etc.), are present within chloroplasts, as they are in mitochondria.

It is not surprising to find, therefore, that isolated chloroplasts can incorporate radio-actively labelled amino acids into their structure. Evidence from these experiments suggests that the radioactivity is located in both the membranes (lamellae), parts of the isolated chloroplasts and the soluble fraction (stroma) (see Fig. 19 on p. 308). In this respect CDNA appears to be far less specific than MDNA, that is, the proteins it codes for are not restricted to the inner membrane alone, but may involve a soluble enzyme of the stroma such a ribulose diphosphate carboxylase. However, although there is evidence that this soluble enzyme is made on chloroplast ribosome, there is no direct evidence that it is coded for by CDNA.

The process of protein synthesis in chloroplasts, however, is made more complex by the fact that as chloroplasts develop from undifferentiated plastids (again see Fig. 19), the process of greening occurs at a specific period in their development. The timing of this process is light-dependent.

Again, although chloramphenicol and erythromycin inhibit protein synthesis in both chloroplasts and mitochondria, cycloheximide shows a variety of effects that depend on the concentration of the antibiotic that is added. Consequently, interpretation of the results of inhibitor studies with chloroplasts has proved very difficult.

7.3.4 Summary of Sections 7.3.2 and 7.3.3

To summarize, then, we can say that both chloroplasts and mitochondria have a considerable degree of autonomy, but they are not completely independent of the nuclear genome. At present, our knowledge about the biogenesis of these organelles is still patchy. It is also important to realize, as in all genetic problems, that the chromosomal and the non-chromosomal genes do not dictate the absolute nature of the finished product. This is modified by environmental factors, which are especi-ally important in chloroplasts and mitochondria. For example, the most important environmental factor in the development of chloroplasts is light. In its absence the internal membranes (grana and lamellae) do not form and synthesis of chlorophyll does not take place. Mineral deficiencies can produce similar effects. In the same way, the development of mitochondria from pro-mitochondria in yeasts is modified by the presence or absence of oxygen and the concentrations of glucose.

7.4 Can the transplantation of cytoplasmic organelles alter the phenotype of the recipient cell?

Chloroplasts and mitochondria are discrete organelles that can be isolated almost free of contaminants from other cell structures. Now that we understand something about the degree of genetic autonomy that they exhibit, it should be possible to design experiments to answer the question posed in the heading.

We have already mentioned the slow-growing respiratory-deficient mutant 'poky' in Section 7.1.2. Other slow-growing mutant strains of *Neurospora* are known; these have been called abnormal 1 (*abn*1) and abnormal 2 (*abn*2), respectively. In both mutants the condition was found to be irreversible and faster rates of growth could not be induced by subculture of the mutants or by the addition of special substrates to the culture medium.

Earlier attempts to see whether injecting cytoplasmic organelles could induce changes in the recipient's phenotype involved micro-injection of the cytoplasm into one or two cells of the wild-type mycelium. After injection, these recipient cells were cut off from the mycelium and grown separately on nutrient agar. Only in a few cases was there an observed change in the phenotype of the recipient. The results were ambiguous for two reasons: first, because nuclear contamination is difficult to control and, second, because the number of mitochondria transferred was comparatively small.

QUESTION How could these two problems be overcome?

ANSWER By extracting a larger number of mitochondria from more cells, purifying them on a sucrose density gradient, and then micro-injecting the purified mitochondria into recipient wild-type cells.

In 1965, Diakumarkos and his colleagues performed this experiment using isolated and purified mitochondria from the *abn*1 strain of *Neurospora*. When the mitochondria were injected into the wild-type mycelia, a number of the cultures showed a decrease or loss of cytochrome a activity, as judged by differences in the absorption spectra in first the oxidized and then the reduced state of the cytochrome. These results suggest that the mitochondria transferred by injection from the respiratory-deficient strain carried with them a genetic determinant that affected the respiratory activity of the wild-type host.

However, even micro-injection of purified and 'concentrated' mitochondria is still at best only qualitatively successful.

7.5 Why have cells evolved a mechanism of cytoplasmic inheritance?

We now come to perhaps the most interesting and yet most difficult question to answer. As you will see from the later Units on population genetics and evolution (Units 9–12), non-essential genes are normally quickly lost in the process of selection. We must, therefore, start with the assumption that cytoplasmic genes do play an essential role in eukaryotic cells. This assumption is justified when we examine a wide range of eukaryotic cells: we find very few cells that can survive without organelle DNA, and it is hardly surprising because chloroplasts and mitochondria are fundamentally important in processes for transforming and transducing energy. If a normally photosynthesizing cell is to survive without CDNA, it has to revert and adapt to an alternative energy supply, that is, it has to become heterotrophic. Among extant algae and higher plants, only the single-celled, motile, *Euglena* has been shown to be able to adapt to both photosynthesizing and heterotrophic conditions. Similarly, if MDNA is lost, it leads to a deficiency in aerobic respiration and the organism must survive by generating energy exclusively from the fermentation pathway. A feature of yeasts is that they can adapt to both aerobic and anaerobic conditions; when loss or damage of MDNA occurs, they become 'facultative anaerobes', that is, as you may recall, they can live and grow by obtaining their energy from a glucose source alone (c.f. the petite mutants).

In addition to these examples, we can cite certain highly specialized cells, such as the mammalian red blood cells, that during their short functional period of carrying oxygen in the circulatory system, dispense with both nucleus and internal organelles altogether. Our list of exceptions, is hardly a lengthy one!

Although the essential role of mitochondria and chloroplasts in energy metabolism is reason enough for their existence in eukaryotic cells, it does not explain why they possess separate organelle DNAs. So, let us examine some of the postulated advantages to the cell of possessing cytoplasmic as well as nuclear genomes.

7.5.1 Hypothetical advantages of two separate genomes and protein-synthesizing systems

The first possible advantage is that uniparental transmission of genetic material (usually maternal) limits recombination in organelle genes and so produces an essentially homozygous cytoplasm that maintains clonal uniformity. If this is correct, then what is the evolutionary value of the resulting cytoplasmic homogeneity?

Chloroplasts and mitochondria function as the main agents supplying cellular energy; evolutionary selection, therefore, must surely have operated strongly on the genome of the organelle to provide the most compact and efficient system. When we examine the ultrastructure of these organelles we find the most complex spatial arrangement of electron-transporting components and enzymes anywhere in the cell. Also, with minor differences, these components are virtually identical in their relative proportions (though not in the total quantity) in a vast range of eukaryotic cells. We must assume that major deviations in the organelle genome are undesirable. Nevertheless, although recombination between cytoplasmic genes is very rare under normal conditions, it has not been eliminated (remember the very low frequency of recombination in *Chlamydomonas*).

A second advantage that has been suggested is that a separately located, but not totally autonomous, cytoplasmic genome is essential for regulating the function, and for assembling the structure, of complex organelles.

You have already seen in this Unit that nuclear and organelle DNAs are replicated at different, but specified, times in the cell cycle (e.g. the recombination and segregation experiments in *Chlamydomonas* and Luck's experiments with a choline-requiring mutant of *Neurospora*). Yet, under normal circumstances the two systems (nuclear and organelle) are sufficiently integrated or coupled so that the amount of DNA in the nucleus and organelles is about the same, despite the variation in the size of the organelles and the widely differing energy demands made on different cells. In spite of this close integration of the two systems, we can as it were 'uncouple' them by noting their different behaviour towards antibiotics. Chloramphenicol, for example, will inhibit organelle protein synthesis but not protein synthesis taking place outside the mitochondrion or the chloroplast, whereas cycloheximide has the reverse effect. Also, there are considerable differences in composition and organization between the chromosomes in the nucleus and the naked DNAs of the organelles, which means that the modes of transcriptional control are almost certainly different. There is evidence of this in the inhibitory behaviour of the antibiotic rifampycin, which prevents RNA polymerase activity in organelle systems, but has no effect on nuclear transcription. So, it would appear that under special 'localized' conditions within the cell, the two systems do function, and need to function, with considerable independence.

Let us remind ourselves again about the suggested control of mitochondrial assembly (Fig. 25).

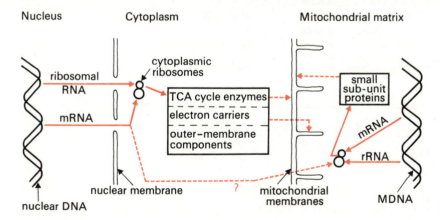

Figure 25 A suggested model for the control of mitochondrial assembly.

Here you can see that some messenger RNAs from the nucleus may enter the organelle for translation on organelle ribosomes.

The idea is attractive, because it means that the organelle ribosomes may have a function of protein synthesis beyond that of just translating organelle genes. However, there are two major objections to this idea. First, there is no experimental evidence to support it. Indeed, what evidence there is suggests that nucleic acids are unable to traverse mitochondrial or chloroplast membranes. Second, if *some* nuclear messenger RNA can be imported into mitochondria or chloroplasts, why not all? Also, if messenger RNA can enter the mitochondrion, why cannot mitochondrial messenger RNA be transported out and 'read' on extra-mitochondrial ribosomes?

However, the second hypothesis put forward above may be less straightforward than it seemed at first sight. Recent biochemical evidence about the relationship between mitochondrial structure and oxidative phosphorylation suggests that the

components of the inner membrane of the mitochondrion are arranged aniso-
tropically, that is, some of the components need to be arranged *directionally* on the
membrane so that they will receive protons and electrons from substrates situated
on the inner side of the membrane only. As a result of dehydrogenase activity (the
removal of hydrogen from the substrates), there is a separation of charge across the
membrane in the form of protons and electrons so that the outer side is more posi-
tively charged than the inner. At specific sites in the membrane where enzymes that
synthesize ATP are present, the separation of charge across the membrane is used
to drive the synthesis of ATP. The directional arrangement of the components and
enzymes is a prerequisite for this mechanism of ATP synthesis. If this hypothesis of
how ATP is made is correct, then two separate genomes with their protein-synthesizing systems could be essential, if the anisotropic arrangement is to be accomplished,
as there is no convincing evidence to support the idea that nucleic acids or large
proteins can be transported right across the inner membranes of these organelles.
Although all the proteins of the outer mitochondrial membrane, and virtually all of
the proteins of the inner membrane and matrix are coded for and synthesized by the
nuclear–ribosomal system, the final positioning and assembly requires the synthesis
of certain small proteins on site. This need is accomplished by the organelle system.
To find out whether this theory is correct or not, we shall have to wait until we know
more about the enzymes synthesized by mitochondria and the precise arrangement
of all proteins in the inner membrane.

7.5.2   The possible origin of chloroplasts and mitochondria via endosymbiosis

In 1890, Altmann suggested that mitochondria were primitive, self-replicating,
bodies within the cell and that they bore many resemblances to bacteria. Later,
in 1922, Wallin suggested that mitochondria 'may be regarded as symbiotic bacteria
whose association with other cytoplasmic components may have arisen in the
earliest stages of evolution ...'.

These speculative ideas may appear to be quite fantastic, but in the light of recent
findings about mitochondria and chloroplasts, we must give serious consideration
to the possibility that both organelles may have originated within host cells many
millions of years ago by endosymbiosis and subsequently may have evolved to their
present state (Sagan, 1967). Several times in this Unit, we have mentioned the similarities that exist between present-day bacteria and mitochondria in certain aspects of
their protein-synthesizing machinery. Similarly, if we examine the structure and
function of primitive blue-green algae, we find many resemblances to chloroplasts.

At this point, therefore, let us just remind you about the similarities between bacteria
and mitochondria.

1   Both possess DNA in a circular form with no associated nucleoproteins (histones).

2   The length of some mitochondrial DNAs (e.g. yeast, which has circular DNA of
some 25 µm in circumferance) are similar to those of bacteria; those of higher organisms (5 µm circles) are much smaller. (Note: The circumferance of chloroplast DNA
is usually longer and in some algae, e.g. *Acetabularia*, may reach 400 µm.)

3   Mitochondrial DNA is self-replicating and lies in the central matrix of the
mitochondrion, which is analogous to the nucleoid region of bacteria.

4   The sequence of bases in MDNA shows a much greater similarity to bacterial
DNA than to nuclear DNA.

5   The sedimentation characteristics (mainly dependent on size) of mitochondrial
ribosomes bear a much closer resemblance to bacterial ribosomes (70 S), than
they do to cytoplasmic ribosomes (80 S). For example, mitochondrial ribosomes
from fungi have sedimentation coefficients between 73 and 74 S; those from higher
organisms are distinctly smaller (between 55 and 60 S).

6   In the sensitivity of the various protein-synthesizing systems to antibiotics there
are greater similarities between mitochondrial and bacterial systems than between
mitochondrial and cytoplasmic systems (Table 3, *opposite*).

Furthermore, in considering the possible evolutionary origin of these organelles, we
might ask why bacteria do not possess mitochondria. An examination of the overall
dimensions of mitochondria and bacteria shows them to be very similar. Indeed, we

Table 3

Antibiotics	Protein-synthesizing systems		
	cytoplasmic	mitochondrial	bacterial
puromycin	+	+	+
cycloheximide	+	−	−
chloramphenicol	−	+	+
erythromycin	−	+	+
lincomycin	−	+	+

+ inhibition   − no inhibition

can speculate further and say that the rod-shaped bacteria resemble many types of mitochondria in shape. If bacteria do not possess mitochondria, how do they respire aerobically? Although the outer capsule of the bacterial cell wall is clearly different from the outer mitochondrial membrane, the inner membrane of the mitochondrion, which houses the components of the respiratory chain (cytochromes, etc.), bears many resemblances to the bacterial cell membrane, which carries the respiratory assemblies. In many bacteria this membrane may be infolded to form a structure called the mesosome. It is thought that in the pre-Cambrian period of geological history, the earth's atmosphere gradually changed from a reducing atmosphere to an oxidizing one, probably brought about by photo-dissociation of water vapour in the atmosphere. This change acted as a selective pressure, resulting in the evolution of porphyrin-containing proteins (i.e. chlorophyll-like molecules). With the gradual onset of photosynthesis among several prokaryotes, the atmosphere came to contain more and more oxygen. As a result, aerobic respiration evolved in prokaryotes. Remember, however, that these are postulated events, not proven facts!

QUESTION If endosymbiosis between certain aerobic bacteria and a larger prokaryotic host cell did indeed occur during the transition from a reducing to an oxidizing environment, can you suggest: (a) what advantages would accrue to the host cell and (b) what advantages would accrue to the invading bacterium?

ANSWER (a) The advantage for the host cell would lie in being able to respire aerobically (provided, of course, that the invading bacteria could be induced to give up some or all of its ATP).

(b) The advantage to the bacterium was presumably the protection afforded by the larger host cell. However, it is also possible that the invading cell suffered some respiratory deficiency in its glycolytic pathway (i.e. the break-down of glucose to lactate or pyruvate), and that this could be provided by the host cell. If endosymbiosis rather than parasitism is to occur, then 'give and take' of this kind would have been necessary.

Among present-day organisms examples can be found that set a possible precedent. In some families of bacteria (the Rickettsiae) that parasitize insect cells, the host is necessary to provide certain essential glycolytic substrates, so that oxidative phosphorylation can take place in the parasite. The parasite, however, does not give up its ATP! There are, therefore, several difficulties with the endosymbiosis hypothesis. First, there is the considerable readjustment in the way of life that would be necessary if these invading organisms were to give up most, if not all, of their ATP to the host cell and, second, because more than one species of bacterium or blue-green alga might be expected to be an 'infective' symbiont, we might expect much greater heterogeneity in chloroplast and mitochondrial populations than appears to occur.

Nevertheless, there are today endosymbionts that have become so remarkably adapted to life with their hosts, that geneticists have had considerable difficulty in deciding whether these are real examples of cytoplasmic inheritance. Perhaps the best known example is 'kappa' inheritance in *Paramecium*, which in recent years has been shown to be caused by a bacterial endosymbiont.

Paramecium aurelia is a free-swimming, unicellular ciliate belonging to the group of eukaryotic organisms known as the Protozoa. Some strains of this species, referred to as 'killers' produce a toxin that is lethal to other strains, referred to as 'sensitives'. Killers release toxic particles from their cytoplasm into the surrounding watery medium and these toxins may be taken up by other *Paramecia* via their gullets and food vacuoles. Evidence that this is the route of entry is shown by gulletless *Paramecia* and by conjugating *Paramecia*, which have non-functioning gullets during the conjugation period. The toxins are produced by kappa particles in the cytoplasm of killers. These particles are about 500 nm in size, contain DNA and RNA and are mutable. The physical properties of this DNA can be readily distinguished from both mitochondrial and nuclear DNA. Kappa particles can be seen with the light microscope using appropriate stains (Fig. 26(a)) in the intact animal, and as isolated components, using phase-contrast microscopy. With phase-contrast (as in Fig. 26(b)) some of the kappa particles (about 20 per cent) show refractile or bright bodies. The toxic particles released by killer *Paramecia* into the medium are whole, bright kappa particles, according to Preer (1974). Under the electron microscope, bright kappa particles consist of a small amount of cytoplasm enclosed within a double membrane (Fig. 26(c)). The appearance (Fig. 26(d)) is very similar to that of *E. coli* shown in Unit 2 in Figure 7(a) on p. 63).

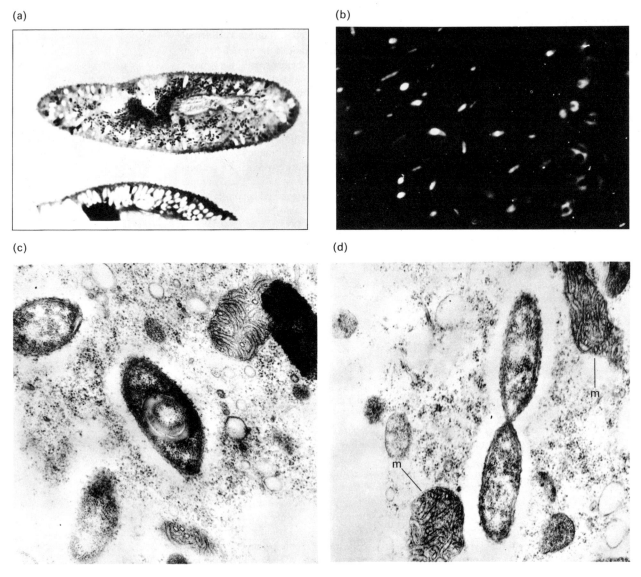

(a)

(b)

(c)

(d)

The presence of kappa in *Paramecium* is dependent on the nuclear gene *K*. Animals of nuclear genotype *kk* are unable to harbour kappa, so kappa is only perpetuated in *Paramecia* of genotypes *KK* and *Kk*. As might be expected killers are resistant to their own toxin. However, different forms of the killer trait exist (e.g. 'spinner' and 'hump' killers) that will kill each other.

In order to understand the inheritance of the killer character, examine the life cycle of *Paramecium aurelia* in your *Life Cycles* folder.

Paramecium is diploid and contains a macronucleus and two micronuclei. Conjugation occurs between cells of different mating types as shown in *Life Cycles*. Pay particular attention to the micronucleus as the macronucleus does not take part in the conjugation process. Instead, a new macronucleus will develop from one of the products of the diploid nucleus formed by conjugation.

Figure 26 (a) Kappa particles within a whole *Paramecium* as seen under the light microscope (stained with osmium lacto-orcein).

(b) Dark phase contrast showing isolated bright kappa particles.

(c) Electron micrograph of a longitudinal section through a bright kappa particle within *Paramecium*.

(d) Electron micrograph showing a non-bright kappa particle in the process of dividing (m = mitochondrion).

As a result of conjugation, exconjugants have heterozygous identical nuclei, but usually differ in their cytoplasm, unless there has been considerable cytoplasmic exchange. Under special conditions*, it is possible to cross killer and sensitive strains of compatible mating types. The resulting clones are heterozygous for K, but cytoplasmically either killers or sensitives. Figure 27(a) shows the normal situation in which there is very little cytoplasmic exchange and (b) shows a rare situation discovered by Sonneborn in America, in which conjugation has been prolonged and considerable cytoplasmic exchange has occurred.

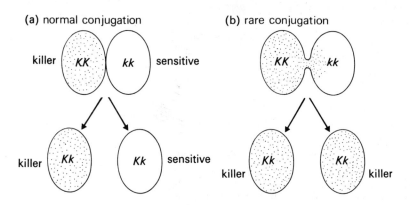

Figure 27 Normal and rare conjugation between 'killers' and 'sensitives' of *Paramecium*.

QUESTION Although these experiments clearly show that a cytoplasmic element is involved, can you conclude that it is determined by a nuclear gene?

ANSWER No.

QUESTION Can you suggest how it might be possible to show that K is required to maintain kappa? (*Hint*: Start with the heterozygous killer cell.)

ANSWER By segregating out the homozygous kk from the cross between Kk and Kk, both of which are killers. If K is required to maintain the killer trait then kappa should be lost after several fissions.

There is, in fact, another way of showing the nuclear-gene requirement for the maintenance of kappa. *Paramecium* can be induced to undergo *autogamy*, that is, the micronucleus undergoes meiosis without conjugation, giving rise to one haploid meiotic product, which divides once mitotically. The two haploid nuclei produced within the same cell then fuse to restore the diploid state.

autogamy

ITQ 4 If the *Paramecium* was heterozygous for the nuclear gene (i.e. Kk) before autogamy, would the diploid cell resulting from autogamy be heterozygous or homozygous? Give reasons for your answer.

ITQ 5 If heterozygous killers (Kk) and heterozygous sensitives (Kk) both underwent autogamy, what proportion of the clones produced would be killers and what proportion would be sensitives?

The results given in the answers to ITQs 4 and 5 demonstrate the classical pattern of cytoplasmic inheritance. Although only *Paramecia* of genotypes KK or Kk *can* harbour and maintain kappa particles, it does not automatically imply that they *will*. Also the experiments show that *Paramecia* can survive and multiply with or without kappa particles, although there is an obvious selective advantage conferred on those animals that do harbour the endosymbiont. The situation is, however, very different from that obtaining for mitochondria and chloroplasts; with very few exceptions the 'endosymbiosis' of chloroplasts and mitochondria is absolutely obligatory, as eukaryotic cells cannot survive without them.

In a review (1974), Preer has collated convincing evidence that kappa and many other endosymbiotic particles (such as 'lambda', 'mu', 'nu', 'pi' and 'sigma') are indeed highly specialized bacteria. For example, the cytochromes of kappa and mu are very similar to those found in bacteria. Ribosomal RNAs from these particles can

* The special conditions are that the two types are reared separately, extensively fed and, when feeding has stopped, pairs are mated in a cavity slide. Once conjugation has finished, the exconjugants are isolated again and the clones are grown separately.

hybridize with DNA from *E. coli* but not with the nuclear DNA of *Paramecium*. Mu and lambda (but not kappa) can apparently be cultured with difficulty, free from the host. Nevertheless, *isolated* kappa particles can respire aerobically using glucose as an energy source because they possess all the necessary enzymes and components. In this respect, therefore, the kappa endosymbiont is quite different from the mitochondrion which possesses only the enzymes of the tricarboxylic-acid cycle and the components (e.g. cytochromes) of the respiratory chain.

The literature of cytoplasmic inheritance contains several examples that can now be attributed to either an endosymbiotic bacterium (e.g. kappa in *Paramecium*), or a virus. Concerning viruses, we might briefly mention the inheritance of carbon-dioxide sensitivity in *Drosophila*. Most insects can be temporarily anaesthetized by exposure to pure CO_2 for a minute or so, but normally recover quickly with no ill effects; as a result CO_2 is frequently preferred to ether by some geneticists when very brief anaesthesia is required. It was of considerable interest, therefore, when in 1958 a strain of *D. melanogaster* was found with a high degree of sensitivity to increased concentrations of carbon dioxide—so much so, that even brief exposures to low concentrations brought about unco-ordinated responses. The trait is transmitted primarily through the maternal parent because the egg contains a larger amount of cytoplasm. Occasionally, however, it is also transmitted paternally via the sperm. Sensitivity may also be induced by micro-injection of cell-free extract from the cytoplasm of flies that are sensitive to CO_2. Extensive tests, together with evidence from electron microscopy have disclosed that the sensitivity is dependent upon an infective virus-like particle in the cytoplasm, called sigma. Endosymbiotic virus-like particles appear to be fairly common in the cytoplasm of insect cells, and some of these particles (but not those with CO_2 sensitivity) may play a significant role in the survival and evolution of insects.

Despite the fact that we can attribute these cytoplasmic characters to either bacteria or viruses, we cannot categorically say that they have nothing to do with cytoplasmic inheritance. As we mentioned in the Introduction to this Unit, it is very difficult to draw a firm distinction between heredity and infection or 'normality' and 'abnormality' because the life cycles of endosymbiont and host are so intricately interwoven. Indeed, we can argue that provided the character (once introduced) is maintained from generation to generation in the cytoplasm and has a nuclear gene requirement for its maintenance, then inheritance must be regarded as cytoplasmic.

7.5.3 Alternative mechanisms that might explain the evolutionary origin of chloroplasts and mitochondria

Despite the present popularity of the endosymbiont theory described in the last Section, there are still several difficulties. Most biologists agree that eukaryotic cells must have evolved from prokaryotic cells, but there are other ways to account for the origin of mitochondria in prokaryotic cells. A feasible alternative was postulated by Mahler and his colleagues in 1972. Their argument is based on the fact that bacteria harbour plasmids. (You have already encountered plasmids in the form of F factors in Units 3 and 4, Section 4.2.4. Plasmids are also more generally referred to by some authors as episomes. In our context, the terms are more or less interchangeable.) Plasmids are carriers of extrachromosomal genetic material in the form of circular DNA molecules, and several plasmids are approximately the same size as many MDNAs. The replication of the chromosomal and plasmid DNA in bacteria is intimately associated with and dependent on specialized structures in the bacterial plasma membrane. Bacteria (prokaryotes) do not contain any membrane-bound organelles, so their respiratory functions are intimately associated with the plasma membrane. Under conditions of physiological stress, however, this membrane can be elaboratory infolded as the drawing from an electron micrograph (Fig. 28(a)) shows.

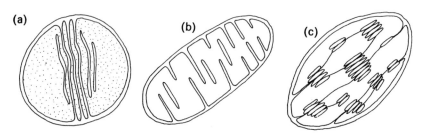

Figure 28 Drawings from electron micrographs showing sectional views of (a) a marine bacterium, (b) a mitochondrion and (c) a higher plant chloroplast.

It is not too difficult to see that such an arrangement of membranes mimics the arrangement of mitochondrial cristae (Fig. 28(b) or the chloroplast lamellae (Fig. 28(c)).

You may also recall from Units 3 and 4, that plasmid (episome) DNA can either exist and replicate independently of the bacterial chromosome DNA, or undergo cycles of integration (via recombination) and detachment (via transduction) from the latter. During this process, it is quite possible that genes could be detached from the bacterial chromosome and become part of the plasmid genome, thereby becoming capable of a separate existence (see Fig. 29).

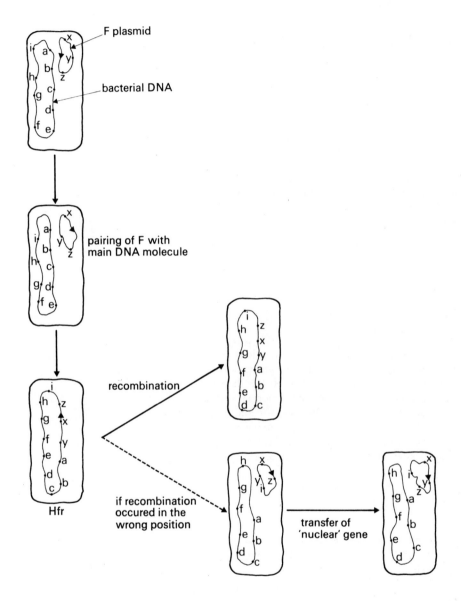

Figure 29 Possible mechanism to explain the origin of MDNA.

Transcription and translation in prokaryotes are closely coupled events, and the necessary machinery for the two processes appears to be always in close association with DNA, so that if a plasmid with some additional information about the respiratory function of the plasma membrane were to become entrapped within a complex membrane structure, we would have basically satisfied the design of a mitochondrion. If some modification of the infolded membrane came about that gave greater efficiency in energy transduction, then the descendants of this cell might well have a distinct evolutionary advantage.

The number of plasmids in certain cells may be as high as 30, and plasmids, like mitochondria, are subject to environmental influences (see Section 7.3.4.)

On the basis of the endosymbiont theory, it is difficult to explain why the symbiont has relegated virtually all of its protein synthesis to the nucleus. Even the striking resemblances between prokaryotes and organelles need not necessarily imply an endosymbiotic origin for these organelles. It can be argued that all three systems have evolved from a common ancestral form. But whereas the protein-synthesizing systems of eukaryotes and their organelles initially evolved together within the same cell from the ancestral form, along a pathway that diverged from that of prokaryote evolution, subsequent evolution of the eukaryotic protein-synthesizing systems has gone on much faster than the organelle system, on the one hand, and the prokaryotic system on the other. Consequently, both organelle and prokaryote systems remain much closer to the ancestral form, that is,

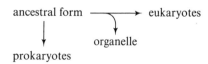

If the plasmid theory is correct, then the multiplicity of plasmids may safeguard such vital functions as respiration and photosynthesis against mutations, in that selection can eliminate the defective plasmid without eliminating the entire cell.

But there are problems inherent in the plasmid theory. For example, it is difficult to perceive that one fragment of DNA associated with a plasmid would contain the genes for mitochondrial ribosomal components, tRNAs and several elements of the inner membrane all together, unless multiple insertions and excisions had taken place between the episome and the DNA of the host cell. There is no known precedent for this. Further criticisms of both the endosymbiotic and plasmid (episomal) theories have been discussed recently by Reijnders (1975), who has himself put forward an alternative hypothesis to account for the origin of mitochondria. If you are interested in persuing this subject further, you are recommended to read the paper which is listed in the Bibliography and References at the back of this Unit.

Whichever theory is correct (if any), speculation about the origin of cytoplasmic organelles is likely to be a subject of argument for many years to come.

7.6   Conclusions

In this Unit we have tried to provide answers to some of the key questions about cytoplasmic inheritance. What emerges clearly, however, is that although the patterns of inheritance as reflected by breeding analyses may seem changed (i.e. they are non-Mendelian), the basic principles of storage of genetic information in nucleic acids and the general concepts of gene replication, gene action and gene mutation have all been reinforced by these studies. There are still, however, many cases of apparent cytoplasmic inheritance, which we have not mentioned in this Unit, whose cellular control mechanisms are even less well understood. An example is provided by the snail, *Limnea*.

7.6.1   Shell-coiling in Limnea

In Figure 30 we show *dextral* (right-handed) and *sinistral* (left-handed) *coiling* of the shell of the snail *Limnea*. The internal organs of the snail adopt a coiling pattern corresponding to the coiling of the shell.

dextral–sinistral coiling

dextral   sinistral

Figure 30 Dextral and sinistral coiling of the shell in the snail *Limnea*.

It seems that the direction of coiling of the shell is determined by a single nuclear gene acting through cytoplasmically produced effects. The dominant allele (s^+) determines that the shell coils in the dextral direction; the allele determining sinistral coiling (s) is recessive.

In a cross between two different homozygous parents, in which the female gamete is (s^+s^+) and the male gamete is (ss), the heterozygous progeny (as expected) are all dextrally coiled. By good fortune, the snails are hermaphrodite and because of this self-fertilization is possible.

326

QUESTION What will be the genotypes and phenotypes of the F_2 progeny from the self-fertilized heterozygotes of the F_1?

ANSWER One-quarter s^+s^+, dextrally coiled; one-half s^+s, dextrally coiled; one-quarter ss. You probably predicted that this group would be sinistrally coiled. Surprisingly, they too are *dextrally* coiled!

It appears that expression of the character is dependent on the maternal genotype. This is further borne out by the fact that if the F_2 progeny are self-fertilized, all those with genotype s^+s^+ and s^+s produce only dextrally coiled progeny in the F_3, whereas the F_2 progeny with genotype ss (but dextral phenotype) produce only sinistral progeny in the F_3.

> ITQ 6 On the basis of these observations from experiments, what would be the phenotypes (and proportions) of the F_1 progeny from an original cross
>
> between $ss(\female)$ and $s^+s^+(\male)$ and the phenotypes of the F_2 and F_3 progeny,
>
> if self-fertilization occurs in the F_1 and F_2 generations?

When this maternally inherited phenomenon is investigated further, we find that the direction of coiling depends upon the orientation of the spindle in the first mitotic division of the zygote (see Unit 2, Section 2.4). In turn, the orientation of the spindle is controlled by the genotype of the oocyte from which the egg develops. The (as yet unknown) control mechanism appears to be built into the egg before meiosis and fertilization occur.

7.6.2 Pollen sterility in maize

Much of what we have said about cytoplasmic inheritance may make fascinating reading, but does it have any relevance or value in the area of applied biology and agriculture?

In most of the examples cited so far in this Unit, the answer is no. It is true that variegated plants have been used for ornamental purposes in horticulture and as such have had considerable commercial value. Decorative plants are certainly aesthetically pleasing, but, unlike food crops, are not essential to human survival. However, there is one very important cytoplasmically inherited feature that has been used to great effect in agriculture. It is the phenomenon of *pollen sterility* and we shall end this Unit by looking at its practical value. Many important crops such as maize, barley, wheat, sugar-cane, onions and beet are grown commercially from hybrid seeds. The importance of outbreeding to form hybrids, rather than inbreeding by self-fertilization, is that the hybrid progeny usually grow more vigorously and give higher yields of foodstuff. But, the production of hybrid seeds requires carefully controlled pollination. Maize and barley, for example, are readily self-fertile and it is essential to block self-pollination if hybrid seed-lines are to be maintained. It is obvious, therefore, that seed-lines carrying a pollen-sterility gene are of great practical value in eliminating self-fertilization. For example, before the development of maize hybrid seed-lines carrying a cytoplasmic-sterility gene in the 1950s, it cost growers an uneconomic sum to de-tassel* maize plants. The cost nowadays would be almost prohibitive.

pollen sterility

Maize seed is usually sold commercially as the hybrid from a 'double cross', originally involving four inbred lines.

In Figure 31, the four inbred lines are represented by maize plants, A, B, C, and D. A and C are plants carrying the cytoplasmic factor T (for Texas, where it was first discovered), which induces pollen sterility; consequently, these plants must be pollinated by another inbred line in which the pollen is fertile, that is, by B or D, respectively. As it happens, pollen fertility can be restored if a dominant nuclear gene (Rf) is present, although its recessive allele (rf) has no effect. As you can see, the F_1 hybrids both carry the cytoplasmic factor T, but whereas hybrid AB plants have sterile pollen, hybrid plants CD have fertile pollen because they carry the nuclear allele Rf. In a cross between AB and CD, AB must be pollinated by CD, although CD is capable of self-fertilization as well (remember pollen is produced very abundantly).

---

* The tassel is the pollen-producing (male) part of the maize plant (see *Life Cycles*).

327

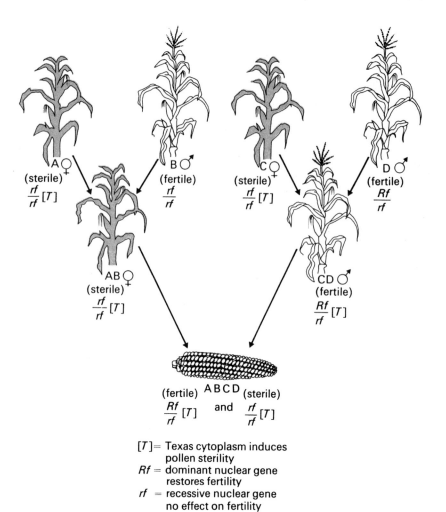

[T] = Texas cytoplasm induces pollen sterility
Rf = dominant nuclear gene restores fertility
rf = recessive nuclear gene no effect on fertility

Figure 31 Production of commercial hybrid maize seeds.

The cobs produced in the F_2 off the AB hybrid parent plant are 'double-cross' hybrids ABCD. Half of the seeds in these cobs will be 'male'–sterile and half will be 'male'–fertile. When these seeds are planted, there will still be ample pollen shed from the male–fertile plants produced, for self-fertilization and cross-fertilization.

Unfortunately, in recent years, seed-growers have been worried by the increase in the incidence of leaf blight (a fungal disease) in maize, especially as the pollen-sterile lines are particularly susceptible to the disease, which is highly destructive. It is to be hoped that this problem will bring about greater interest in a more fundamental understanding of cytoplasmic inheritance in general and crops such as maize in particular.

Self-assessment questions

Section 7.1

SAQ 1 Most strains of *Chlamydomonas* are sensitive to streptomycin (*ss*). If you found a strain that was streptomycin-dependent (*sd*), how would you decide whether the character was determined by a nuclear gene or a cytoplasmic gene?

SAQ 2 Slow growth in *Neurospora crassa* is frequently caused by a cytoplasmic gene that gives rise to a respiratory deficiency. There are a number of different cytoplasmic genes that induce this effect, of which the poky mutant is one, and they are all transmitted through the protoperithecium of one parent. Certain nuclear genes are known to increase the rate of growth, even when a respiratory deficiency is present. One gene that promotes fast growth in the respiratory-deficient mutant, but has no effect on the wild type, is the dominant nuclear gene *F*, whose allele *f* does not promote growth under any conditions.

If the respiratory-deficient protoperithecial parent carrying the *F* gene is crossed with the wild-type conidial parent carrying the *f* allele, what would be the expected genotypes and phenotypes of the resulting ascospores if the respiratory deficiency is not permanently modified by nuclear gene *F*?

SAQ 3 Jinks in 1963 developed a 'heterokaryon test' in the fungus, *Aspergillus nidulans* (see *Life Cycles*), to provide evidence, in the absence of a sexual stage in the life cycle, of whether a given phenotype was determined by a nuclear gene or not.

A heterokaryon was made between the mycelia of two compatible haploid strains of fungi. One of the strains produced green spores, formed diffuse colonies on nutrient agar and had an unpigmented mycelium. The other strain produced white spores, formed compact colonies on nutrient agar and had a mycelium with a purple pigmentation, that is,

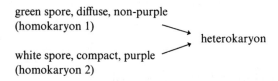

green spore, diffuse, non-purple
(homokaryon 1)

white spore, compact, purple
(homokaryon 2)

→ heterokaryon

The heterokaryon gives rise to *haploid* spores that are *asexually* produced. 400 of these haploid spores were analysed by growing each separately on agar plates:

147 spores gave rise to green, diffuse, non-purple phenotypes,

121 spores gave rise to white, compact, purple phenotypes,

103 spores gave rise to green, diffuse, purple phenotypes, and

 29 spores gave rise to white, compact, non-purple phenotypes.

Which of the characters are likely to be determined by nuclear genes and which by cytoplasmic genes? Give reasons for your decisions.

Section 7.5

SAQ 4 If conjugation between the following mating types ((i)–(vii)) of *Paramecium aurelia* were to take place, predict the percentages of the exconjugants from (a) normal conjugation and (b) rare matings where cytoplasmic mixing occurs that will possess and then be able to maintain kappa particles through several asexually produced generations.

(i)   *KK* (kappa present) × *KK* (kappa absent)

(ii)   *KK* (kappa present) × *Kk* (kappa absent)

(iii)   *KK* (kappa present) × *kk* (kappa absent)

(iv)   *Kk* (kappa present) × *Kk* (kappa absent)

(v)   *Kk* (kappa present) × *kk* (kappa absent)

(vi)   *kk* (kappa present) × *kk* (kappa absent)

(vii) *Kk* (kappa present) autogamy

329

Section 7.6

SAQ 5 The production of hybrid seeds in several cereal crops, particularly in maize and barley, demands controlled cross-pollination. It is most important, therefore, to prevent self-pollination, so the use of seed-lines carrying a pollen-sterility gene has been of great practical value.

A number of male–sterile maize plants are known that are determined by either nuclear genes or cytoplasmic genes.

(a) Explain why all known male–sterile genes are recessive.

(b) How is it possible to distinguish between a cytoplasmic gene for pollen sterility and a nuclear gene for pollen sterility?

(c) If a male–sterile plant that is determined by a nuclear gene is pollinated by a pure-breeding normal wild-type plant and the F_1 is interbred, what phenotypes (if any) will be observed in the F_2 progeny?

(d) If a cytoplasmically-determined male–sterile plant is pollinated by a pure-breeding normal wild-type plant and the F_1 is interbred, what phenotypes (if any) will be observed in the F_2 progeny?

General

SAQ 6 Here are a number of assertions, each followed by a reason.
State whether the assertion and then the reason in each case is *true* or *false*; If true, state whether the reason follows on from the assertion.

(i) *Assertion* The petite phenotype in yeast manifests itself as small, slow-growing, colonies on minimal medium containing glycerol and low concentrations of glucose.

Reason because petites are unable to grow on non-fermentable carbon sources.

(ii) *Assertion* The degree of suppressiveness in certain petite mutants in yeast is never 100 per cent

Reason because at least some of the diploid cells on fusion must be 'normal' in order to sporulate.

(iii) *Assertion* Cytoplasmic genes have been identified in *Chlamydomonas* and distinguished from nuclear genes on the basis of genetic analysis

Reason because the cytoplasmic genes do not segregate at meiosis, but during mitosis of each zygospore clone.

(iv) *Assertion* The white leaf areas in variegated iojap maize contain small colourless plastids

Reason because the double recessive homozygous gene (*ij//ij*) is necessary to induce and maintain them.

(v) *Assertion* Mitochondria arise from pre-existing mitochondria by growth and division rather than by *de novo* synthesis

Reason because radioactive-labelling studies show that as the cell mass doubles, the average grain count is halved, but in a randomized pattern of labelling.

(vi) *Assertion* DNA fibres, usually in the form of circular molecules, exist within the matrix of mitochondria

Reason because experiments with DNase and RNase show that both these enzymes remove the fibre-like structures.

(vii) *Assertion* Purified, isolated mitochondria possess their own RNA

Reason because when such mitochondria are centrifuged in a sucrose density gradient, the major RNA activity peak corresponds with the peak of cytochrome oxidase activity.

(viii) *Assertion* Mitochondria and chloroplasts must have evolved from prokaryotes

Reason because chloramphenicol and erythromycin inhibit protein synthesis in these organelles and in prokaryotes, but not in eukaryotes.

Answers to ITQs

ITQ 1 (*Objective 3*) By definition, recombination is the appearance of new combinations of genes in the progeny after the fusion and segregation of two gametes differing in genetic constitution. If only one gamete contributes to the cytoplasmic genotype, recombinants are impossible. Of course, conventional mapping procedures rely on the frequency of recombinants obtained from crosses with different linked genes; as a result mapping, too, would be impossible where uniparental inheritance is involved.

ITQ 2 (*Objective 5*) You may have suggested the following, although the list is by no means exhaustive.

1   Their size and relative molecular masses

2   Whether they are linear or circular molecules

3   Whether they are single or multi-stranded

4   Their base compositions and sequences and, therefore, their coding capacities

5   Their buoyant densities in a suitable gradient mixture.

ITQ 3 (*Objectives 2 and 3*) In a cross between two homozygotes ($ij^+//ij^+$ and $ij//ij$) all of the F_1 should have the same genotype ($ij^+//ij$). This should be true of both crosses and the reciprocal crosses should give identical results. (Note: Even if ij were not a recessive gene, but either dominant or codominant, the three different F_1 phenotypes cannot be explained on a Mendelian basis.)

ITQ 4 (*Objectives 2 and 8*) Homozygous for either K or k. In a population of cells undergoing autogamy, where the meiotic product is randomly 'selected', 50 per cent of the offspring will be of genotype KK and 50 per cent of genotype kk.

ITQ 5 (*Objectives 2 and 8*) All of the heterozygous sensitives that undergo autogamy remain sensitive; but only half the heterozygous killers that undergo autogamy remain killers, reflecting the segregation of the nuclear genes.

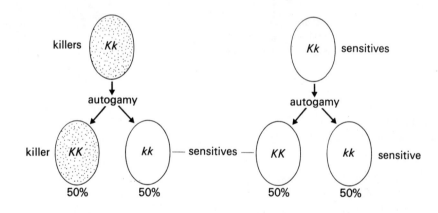

Figure 32

ITQ 6 (*Objective 9*) F_1 *generation*
All progeny will be phenotypically *sinistral* like their mother, although genotypically s^+s. The main determinant is the mother's genotype which was ss.

F_2 *generation*
All progeny will be phenotypically *dextral*. This may have surprised you, but using the same principle as before, the mother's genotype (carrying the dominant allele) s^+s is all important to the next generation.

F_3 *generation*
All progeny whose mothers had the genotype s^+s^+ or s^+s produce only *dextral* progeny. The progeny from mothers with genotype (ss) are all *sinistral*, despite the fact that the mother's phenotype was dextral.

Answers to SAQs

SAQ 1 (*Objectives 2, 3 and 4*) Theoretically, the answer to this question is fairly straightforward. First, it would be necessary to select compatible mating types mt^+ and mt^-, making sure that the mt^+ strain also carried the streptomycin-dependent gene (*sd*) and the mt^- strain the streptomycin-sensitive gene (*ss*).

If these are all unlinked nuclear genes then the progeny will be expected to segregate in a ratio of 1:1:1:1, approximately (one-quarter *sd*, mt^+; one-quarter *sd* mt^-; one-quarter *ss*, mt^+; one-quarter *ss*, mt^-).

If the genes are linked in the nucleus, then a 1:1 ratio of the parental types (i.e. *sd*, mt^+; *ss*, mt^-) will be expected, with perhaps a few recombinant types.

If, however, the character is determined by a cytoplasmic gene, then virtually all the progeny should follow the mt^+ parental line and be streptomycin-dependent, even though fusion is isogamous in *Chlamydomonas*, that is, the ratio would be 4 *sd*:0 *ss*, and the nuclear genes mt^+ and mt^- would segregate in a 1:1 ratio.

The determination of the cytoplasmic character in experiments could be more difficult, because progeny carrying the *sd* gene will grow only on streptomycin agar, whereas *ss* progeny will not and, conversely, *ss* progeny will grow only on minimal agar, but *sd* progeny will not. The way round this problem is to carry out the cross with the motile cells as before, but to then plate out the zygospores produced at known positions on (a) minimal agar to select for *ss* and on (b) agar plus streptomycin to select for *sd*. The number of colonies that grow on both types of agar against the number that do not can be used additively to determine the ratio of the two phenotypes.

SAQ 2 (*Objectives 2, 3, 4 and 5*)

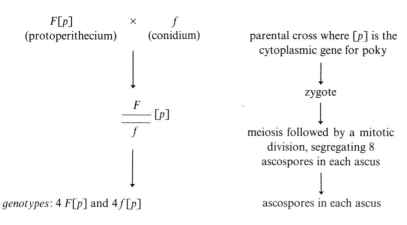

genotypes: 4 $F[p]$ and 4 $f[p]$

phenotypes: 4 ascospores giving colonies showing fast growth; and
4 ascospores exhibiting the slow-growing, respiratory-deficient, poky mutant.

SAQ 3 (*Objectives 2, 3, 4 and 5*) It is stressed in the question that the heterokaryon produced from the two strains of *Aspergillus nidulans* maintains the two genetically distinct types of haploid nuclei and that it also gives rise to asexually produced haploid spores. At no stage, therefore, is there nuclear fusion followed by recombination and segregation of genes. This being so, it would appear from the data that the purple phenotype is cytoplasmically inherited, whereas spore colour and colony shape are determined by nuclear genes.

If all three characters were determined by nuclear genes (irrespective of linkage or non-linkage), only two classes of asexual spores would be expected, each resembling one or the other parent.

Although there is no nuclear fusion, there is variable cytoplasmic mixing in the heterokaryon. This could give rise to either one phenotype (i.e. purple) or the other (i.e. non-purple) to a variable degree and this variability will tend to be reflected in the variable colony numbers obtained in each of the four classes. As it happens, the data here do show a preponderance of the two parental types, although the white-spored, compact colony, phenotypes do appear to be at a selective disadvantage for some reason in comparison with their wild-type counterparts.

SAQ 4 (*Objectives 2, 3, 4 and 8*)

| | Mating type | | Percentage of | |
	(kappa present)	(kappa absent)	(a) normal conjugation	(b) rare matings (cytoplasmic mixing)
(i)	KK	× KK	50	100
(ii)	KK	× Kk	50	100
(iii)	KK	× kk	50	100
(iv)	Kk	× Kk	50	75
(v)	Kk	× kk	25	50
(vi)	kk	× kk	0	0
(vii)	Kk (autogamy)	—	50	—

SAQ 5 (*Objectives 2, 3, 4 and 10*)

(a)   Self-fertilization would be impossible if the male–sterile gene was dominant; only cross-pollination by a fertile plant would be possible. It is highly probable, therefore, that the dominant gene would be eliminated from heterozygotes within relatively few generations by continuous backcrossing to normal fertile-pollen plants.

(b)   Characteristically, the male–sterile cytoplasmic gene exhibits strict maternal inheritance. Consequently, all the F_1 progeny would be male–sterile (i.e. the pollen they in turn produce would be useless). If a recessive nuclear gene for male sterility were involved, then a cross between this plant and normal pollen would yield all phenotypically normal plants (i.e. the pollen produced by the F_1 progeny would be fertile).

(c)   As the F_1 will be phenotypically normal, but heterozygous (genotype $ms^+//ms$), 75 per cent of the F_2 will be fertile (i.e. one-quarter $ms^+//ms^+$; one-half $ms^+//ms$), and 25 per cent will be male–sterile (i.e. one-quarter $ms//ms$).

(d)   An F_2 generation cannot be produced as none of the F_1 plants produce fertile pollen.

SAQ 6 (*Objectives 4, 5, 6 and 7*)

(i) *Assertion* True.

Reason True. But it does not logically follow from the assertion. They are both quite different statements about the petite phenotype.

(ii) *Assertion* True.

Reason True. This is certainly one but not the only reason for the degree of suppressiveness.

(iii) *Assertion* True.

Reason True. The reason given logically and correctly follows from the assertion.

(iv) *Assertion* True.

Reason False. The statement is correct in so far as $ij//ij$ is necessary to induce the variegated (iojap) condition (although it does not induce a change in all plastids). Nevertheless, once this change is initiated, it is irreversible and inherited through the maternal cytoplasm irrespective of the genotype of the male; $ij//ij$ is not, therefore, necessary for the continued maintenance of the variegated (iojap) phenotype.

(v) *Assertion* True in virtually all cases.

Reason True. The reason is the experimental evidence supporting the assertion.

(vi) *Assertion* True.

Reason False. RNase has no effect on the fibre-like structures and DNase shows an effect only on thin sections of mitochondria or on isolated MDNA. It has virtually no effect on MDNA within intact mitochondria.

(vii) *Assertion* True.

Reason True. The reason does logically follow because, once again, this is experimental evidence in support of the assertion.

(viii) *Assertion* Probably true in so far as eukaryotes (and their organelles) evolved from prokaryotes, although the word 'must' is rather emphatic.

Reason True. But it does not necessarily follow from the assertion. This tentative evidence can be interpreted in another way as given in the text (on p. 326).

Bibliography and references

1 General reading

Racker, E. (ed.) (1970) *Membranes of Mitochondria and Chloroplasts*, A. C. S. Monograph 165, Van Nostrand-Reinhold. (Chapter 7, by G. Schatz is useful background reading.)

Sager, R. (1972) *Cytoplasmic Genes and Organelles*, Academic Press.

Jinks, J. L. (1964) *Extrachromosomal Inheritance*, Prentice-Hall.

Jinks, J. L. (1976) *Cytoplasmic Inheritance*, Oxford Biology Reader No. 72, Oxford University Press.

Roodyn, D. B. and Wilkie, D. (1968) *The Biogenesis of Mitochondria*, Methuen Monographs.

Tribe, M. A. and Whittaker, P. W. (1972) *Chloroplasts and Mitochondria*, Studies in Biology No. 31, Edward Arnold.

2 Publications cited in the text

Preer, J. R., Preer, L. V. and Jurand, A. (1974) Kappa and other endosymbionts in *Paramecium aurelia*, *Bact. Rev.*, **38**, No. 2, 113–63.

Raff, R. A. and Mahler, H. R. (1972) The non-symbiotic origin of mitochondria, *Proc. Nat. Acad. Sci.*, **177**, 575–82.

Reijnders, L. (1975) The origin of mitochondria, *J. Mol. Evol.*, **5**, 167–72.

Sagan, L. (1967) On the origin of mitosing cells, *J. Theoret. Biol.*, **14**, 225–42.

Acknowledgements

Grateful acknowledgement is made to the following for material used in this Unit:

Figures 5, 7–13, 18, 24 and 31 adapted from Professor Ruth Sager, *Cytoplasmic Genes and Organelles*, Academic Press, 1972; *Figures 14, 15 and 16* from M. M. K. Nass and S. Nass, *Journal of Cell Biology*, Rockefeller University Press, 1963; *Figure 17* from M. M. K. Nass, *Proceedings of the National Academy of Sciences*, **56**, 1966; *Figure 19* after Von Wettstein; *Figure 21* after Barnett and Brown, from D. Roodyn and D. Wilkie, *The Biogenesis of Mitochondria*, Methuen; *Figure 22(a)* by courtesy of Professor K. Porter, from M. Tribe and P. Whittaker, *Chloroplasts and Mitochondria*, Edward Arnold, 1972, reprinted 1974; *Figures 22(b), 23 and 25* from M. Tribe and P. Whittaker, *Chloroplasts and Mitochondria*, Edward Arnold, 1972, reprinted 1974; *Figure 26(a)* from A. Jurand and G. G. Selman, *The Anatomy of Paramecium aurelia*, Macmillan, 1969, by permission of Macmillan and Co., London and Basingstoke; *Figures 26(b–d)* by courtesy of Dr. A. Jurand, Department of Genetics, University of Edinburgh.

8 Developmental Genetics

Contents

List of scientific terms used in Unit 8

Developed in this Unit	Page No.	Developed in this Unit	Page No.
blastula	338	homoeotic mutations	355
cell-autonomous (gene action)	354	imaginal disc	355
		intra-uterine environmental effect	364
cleavage	343		
clone	354	lethal phase	346
determined cell	357	mitotic crossing over	352
differentiated cell	339	oogenesis	363
eversion	369	phenocopy	369
fate map	360	phenocritical period	346
focus of a mutation	352	pleiotropy	366
genetic chimaera	354	poleplasm	362
genetic composite	352	temperature-sensitive lethal mutation	348
genetic maternal effect	363		
genetic mosaic	352	temperature-effective period	379
gynander	359		
homeostatic	370	totipotent	342

Objectives for Unit 8

After studying this unit you should be able to:

1 Define, recognize the best definition of, and place in the correct context, the items in the list of scientific terms opposite.
(SAQ 6)

2 Explain, with examples, why it is thought that phenotypic differences that appear between cells during development are the result of the control of gene expression rather than the result of segregation of different genes into different cells.
(ITQs 1 and 6)

3 Give evidence supporting the idea that development is the result of organized variation in gene activity in different cells and at different times.
(ITQs 2, 3, 4 and 7; SAQs 1 and 2)

4 Describe the application of induced genetic composites in the investigation of aspects of genetic activity during development.
(ITQs 4 and 5; SAQ 2)

5 Describe and discuss experiments that demonstrate the reciprocal nature of the interactions between nucleus and cytoplasm in cells during development.
(ITQ 7; SAQ 3)

6 Correlate aspects of the structure and behaviour of chromosomes during development with the hypothesis that development follows from systematic differences in gene activity.
(SAQ 5)

7 Recall and discuss specific examples of pleiotropy in gene expression.
(ITQ 8; SAQ 4)

8 Describe and discuss examples of procedures that geneticists use in examining the concept that the genome represents the programme of instructions for development.
(ITQs 4, 5 and 9)

Study guide for Unit 8

This Unit completes the section of the Course on the basic phenomena of genetics and the biochemical aspects that arise directly from those phenomena. In a sense Unit 8 brings you full circle back to Unit 1, in which you first met the idea of the distinction between genotype and phenotype, and deals with the events of development that result in the formation of individual organisms with various phenotypes. For this reason, the visual aspect is very important (e.g. pictures of features and processes of development and of what various structures such as the insect egg look like in 'close-up'), and you should pay special attention to the related Broadcasts, in particular, TV programmes 7 and 8, and Radio programme 8. Also, from the Home Experiment you will be familiar with some aspects of the appearance of *Drosophila melanogaster*, which features prominently in this Unit.

Read the Unit through in sequence, noting that when you reach Section 8.4.3 you are asked to read the offprint, 'Genetic dissection of behaviour' by S. Benzer, which is included with this Unit as required reading. The ITQs will allow you to test your comprehension of various points; the SAQs at the end of the Unit relate to the Objectives and will help you to check that you have understood them. At the end of each main Section there is a Summary that relates the topics to the Objectives; you may find these useful for revision afterwards.

Certain Sections and topics in this Unit are more important than others (a) in order to achieve the Objectives of this Unit and (b) to complete the rest of the Course satisfactorily. If you are short of time when you start work on the Unit, you may consider Sections 8.4.4, 8.5.1, 8.5.3 and 8.8.1 as optional. If the nuclear transplantation experiment in Section 8.1 is clear, you may omit Section 8.2, or be content to scan it briefly. Section 8.4 describes important and central issues and procedures in developmental genetics. You may omit all of this Section, together with the associated paper by S. Benzer, without impairing your study of subsequent Units, though you would then be unlikely to achieve Objective 4 of this Unit.

8.0 Introduction to Unit 8

The analysis of development is an exciting and yet enigmatic problem in biology. It has attracted both the philosopher in biology and also the molecular biologist fresh from solving the riddle of the genetic code. It is the focus of considerable contemporary study, although discussions and speculations about generation and reproduction have gone on for centuries. Until the present century, development and inheritance were considered together as one problem, as facets of the reproduction of organisms. Rather strangely, with the rise of Mendelism early this century the study of inheritance and the study of development became divorced from one another. Nevertheless, it is probably apparent to you that genetics is initimately related to development! Within this Unit we are going to explore this relationship.

It is fair to point out to you that we have already touched upon development in this Course, although perhaps obliquely. You do know that genes are somehow implicated in development. Why? Because mutations in the genes give rise to phenotypic abnormalities, which themselves are the end-product of development. In fact, at the start of this Course, this was how we recognized and defined genes—as explanations of heritable phenotypic differences (Unit 1). In Unit 6 clear evidence was given that the DNA sequence of a gene codes for the primary sequence of a polypeptide and that many proteins have catalytic functions, that is, they are enzymes. These enzymes in turn exert changes in the molecular composition of the cells in which they are produced. The study of auxotrophic mutants and the biochemistry of the particular biosynthetic pathways grew in parallel, historically; the genetic definition of a series of steps in a biosynthetic pathway was complemented by the biochemical definition of the several enzyme-mediated steps involved in the pathway. What emerged was the concept that each of the steps in a pathway was ultimately controlled by the activity of a particular gene (Unit 6, Section 6.3). Finally, you will have seen TV programme 6 in which the process of specification and assembly of a functioning virus was 'dissected' genetically into a series of related operations, each one reflecting the activity of a particular gene at a specific and critical phase of the assembly. In short, what we are saying is that as genes code for the primary molecular units of biosynthetic processes, it might be reasonable to presume that the process of development itself may be the result of a concert of gene activity in a growing organism.

Now, we do try in this Course to avoid presumptions of this kind and so, although not forgetting this argument, let us set it aside for the moment and instead examine three particular examples in which genes are implicated in developmental processes to see what questions they raise. Then we shall look at the methods available to geneticists to see how genetics is being used to analyse these and similar phenomena.

We do not expect you necessarily to be familiar with any detailed notions of development in particular organisms. Descriptive developmental biology is an enormous body of knowledge and we shall try to restrict descriptive aspects of developmental processes that may be new to you to the minimum needed to understand the points we are making. Bear in mind that we shall be concentrating on two aspects:

1   The relationship between gene action and the process of development.

2   The use of genetic procedures and methods to analyse and draw conclusions about development.

Let us look at three examples of the involvement of genetics in development.

8.1 An outline of three case studies in developmental genetics

Our first example is taken from *Xenopus laevis*, the clawed toad. It is possible to remove the nucleus from a differentiated skin cell of an adult frog, and to inject that nucleus into an egg from which the nucleus has previously been removed or rendered inactive (see Fig. 1).

This is quite a technical feat and what happens is that the 'new' egg containing the nucleus from the skin cell begins to develop into a ball of cells by mitosis, just as any normal zygote would. This stage is called the *blastula* stage of the embryo. If nuclei from such a blastula are then again transplanted to other eggs from which the nuclei

blastula

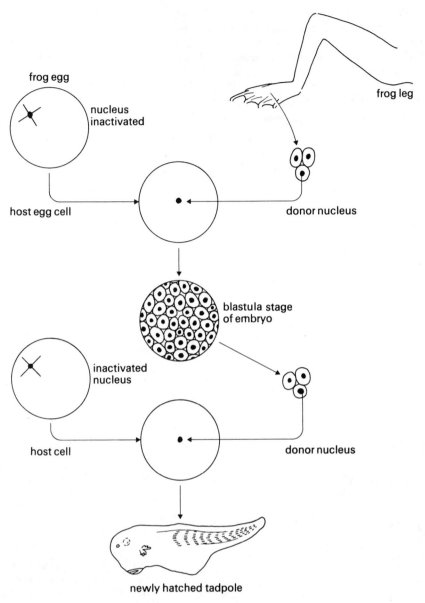

Figure 1 Nuclear transplantation from adult cells to eggs in *Xenopus laevis*.

have been removed, these eggs similarly develop into swimming tadpoles, each with a functioning heart, muscles and nerves. Before it was removed, the nucleus was in a *differentiated* skin cell; that cell type had distinctive biochemical, physiological and morphological characteristics unlike those of the original embryonic cells from which it was derived by a process of differentiation. It is thought that this differentiation process must at some stage involve the production of particular gene products to give the differentiated cell its characteristics. Yet the transplantation experiment of Figure 1 has revealed that the nucleus of the cell still retains the ability to give rise in an ordered way (just as a normal embryo does) to the many diverse cell types of the tadpole.

differentiated cell

ITQ 1 Choose the statement(s) that are the most correct conclusions to be drawn from this experiment.

(i) The nucleus of the skin cell formed that particular skin cell because it contained only the genes necessary for that type of cell.

(ii) The nucleus from the skin cell contained all the genes needed for forming all differentiated cell types and also for the normal progression of development in a tadpole.

(iii) The removal of the nucleus from one cytoplasmic environment to another altered its behaviour—its gene activity changed, resulting in the different cell types and different gene products after cell division.

(iv) The nucleus from the skin cell could not have come from a correctly and completely differentiated cell type because it subsequently gave rise to a complete tadpole after transplantation back into an egg.

The answers to the ITQs are on pp. 377–9.

We conclude from this experiment that cells can differentiate in different ways and yet still retain all the genes necessary for the complete development of a tadpole. Therefore, how can the genes be responsible for the differences we see arising between cells during development? The appearance of different biochemical attributes must reflect differences in the genes that are used or expressed. You might like to contrast this conclusion with the idea held earlier in the century that different cells in an organism were different from each other because they actually received different collections of genes when the cells that formed them went through their divisions.

Our second phenomenon comes from *Drosophila melanogaster*. The phenotype of the X-linked recessive mutation, 'rudimentary', is a reduction in wing size. Figure 2 shows the unusually small and deformed appearance of these wings compared with the wing of a wild-type fly.

(a)

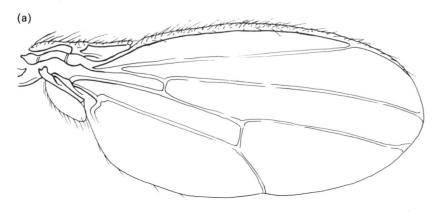

(b)

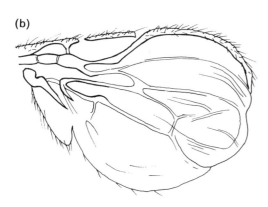

Figure 2 Drawings of wings of *Drosophila melanogaster*: (a) the wild type and (b) a homozygote for the mutation rudimentary wing.

Moreover, females homozygous for rudimentary ($r/ /r$) lay eggs that are not fertile (i.e. they do not develop) unless the sperm fertilizing them brings in a normal allele (r^+) of the gene. We indicate below the results of the crosses on which this conclusion is based.

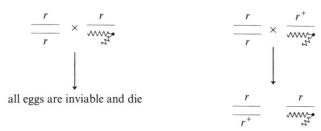

all eggs are inviable and die

all female progeny survive; all male progeny die as embryos

It was recently discovered that the sterility of the homozygous rudimentary females could be cured if extra pyrimidines were added to their diet. Subsequently, biochemical analysis showed that in rudimentary flies three enzymes concerned in the synthesis of pyrimidines were deficient. Thus, a basic metabolic defect—and it is basic because pyrimidines are, of course, used in nucleic acids—was seen to be related to the formation of a normal wing.

The third and final example also comes from *D. melanogaster*. The recessive mutation 'bicaudal', when homozygous in females, causes them to lay defective eggs. Many eggs do not hatch and in some of the eggs one can clearly see that the embryo is abnormal and has developed two abdominal parts in mirror-image relationship to each other, with no sign of the structures typical of the anterior part of the embryo. Figure 3 shows the appearance of a normal embryo at this stage (a) in dorsal view and (b) in ventral view, and (c) shows a defective embryo that, although not actually derived from a bicaudal female, has the appearance characteristic of this mutation.

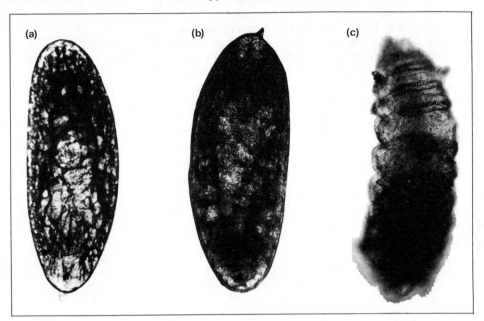

This defect occurs in the eggs irrespective of what genotype the male that has mated with the homozygous bicaudal female carries. Clearly, this mutation has irreversibly upset some aspect of organization in the egg, because even when the normal allele of bicaudal is introduced to the egg by the sperm, the situation is irretrievable.

Figure 3 The bicaudal effect in *D. melanogaster*: (a) dorsal and (b) ventral views of wild-type embryos; (c) a bicaudal embryo at hatching from the egg with two sets of caudal structures joined end-to-end in mirror-image symmetry.

Now that we have three rather striking phenomena in front of us, all implicating genetics in development, let us see what questions do arise. The first experiment with *Xenopus* suggests that differences arising between cells during development may be due to differences in the expression or activity of genes in different cells, rather than to different cells containing different genes and as a result becoming phenotypically different. In the next Section (Section 8.2) we shall examine this question in more detail to see how widespread this finding is and exactly how confident we can be that the interpretation given is the only reasonable one. Is there direct evidence to show that differences between cells can arise from differences in gene expression? It ought to be true that if the expression of particular genes characterizes particular differentiated cells and developmental processes, then mutations in these genes should have specific effects—effects that might be limited to particular cells or tissues, or to particular periods in development when those genes are used. In Section 8.3 we shall be looking at the evidence on this point.

We concluded earlier (see ITQ 1) that the behaviour of a nucleus might be conditioned by the cytoplasmic environment in which it found itself. In addition, the bicaudal effect shows that the cytoplasmic environment of a zygotic nucleus can be upset to such an extent that a normal diploid nucleus carrying the wild-type allele of bicaudal will not give rise to a normal embryo. We can thus ask, what is the nature and the extent of such interaction between nucleus and cytoplasm in the developmental process (the topic of Section 8.5). The mutation rudimentary, on the other hand, forces us to examine another question: what is the relationship between cellular biochemistry and development? Although there is one basic similarity between the expression of rudimentary and of bicaudal—they are both effects exerted by the maternal genotype upon their progeny (eggs)—they may have very different explanations. Remember that the sterility effect of the rudimentary mutation is 'relieved' or cured by pyrimidines added to the diet of the females, although so far no cure has been achieved for the effect of the bicaudal mutation. Are there then different hierarchies in the developmental process (Section 8.6)?

Finally, the stage will be set for us to examine the notion, only a notion at this point and not yet a hypothesis, that perhaps the genes together (the genome, in other words) can be thought of as the instructions for a programme of developmental effects (this is the topic of Section 8.7).

341

8.2 Nuclear transplantation and the concept of totipotency

8.2.1 Totipotency of somatic nuclei from adult frogs

The essential features of the nuclear transplantation procedure that showed that the nucleus of an adult somatic cell could eventually support the development of a swimming tadpole were summarized in Figure 1. When it can be shown that the nucleus of a differentiated cell-type is still able to give rise to other cell types in an organized fashion, as in a swimming tadpole, then that nucleus is said to be *totipotent*; it still has the potential to redifferentiate into other types of cell. Such an experiment as that described in Section 8.1 demonstrates that the nucleus from the skin cell is still totipotent. Or does it? Experimental findings that on face value were similar to those described in Section 8.1 have been in the literature for a number of years and yet this particular experiment with keratinized skin cells was published as recently as 1975 by Gurdon, Laskey and Reeves. There must have been more to this experiment than merely repeating an earlier observation; indeed, this particular experiment is definitive, as we shall see.

totipotent

Gurdon's experiment included a number of precautions designed to eliminate several possible explanations, alternatives to totipotency. The full procedure is illustrated in Figure 4, which summarizes the steps in the experiment.

Figure 4 Steps in the transplantation procedure that demonstrates totipotency in *X. laevis*.

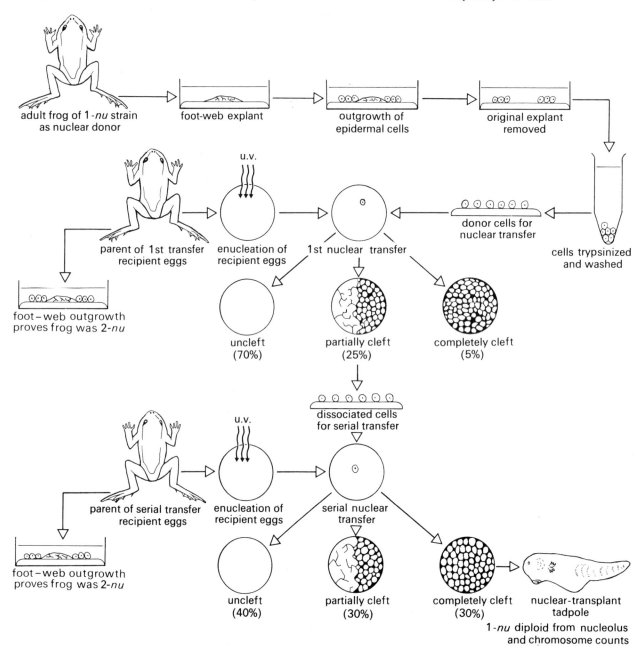

342

Pieces of the foot-web of adult frogs were removed to a culture medium in which cells from the explant, which had now ceased dividing, formed a single-cell layer and were synthesizing keratin. Keratin is a protein commonly formed by 'skin' cells and it contributes to the toughness and impermeability of surface layers of cells in organisms. Keratin production was detected by using a fluorescent antibody prepared against purified keratin from *Xenopus laevis*. If this antibody is applied to cells, and they are examined in ultraviolet illumination, cells that have formed keratin will bind the antibody and show fluorescence. This technique of 'labelling' with fluorescent antibodies showed that after 3 days of culture, 99.9 per cent of the foot-web cells were producing keratin. Nuclei for the transplantations were taken from these cells half a day later. The recipient egg nucleus (the nucleus of the 'host' cell) was inactivated and destroyed by ultraviolet irradiation of the egg before transplantation. The skin cells came from a genetic stock in which all cells had only one nucleolus. For our purposes, you need only know that a nucleolus is a densely staining region associated with the chromosomes, visible in the cells. The recipient egg was taken from a stock in which each diploid nucleus had two nucleoli. As Figure 4 shows, after the first nuclear transfer, only 30 per cent of the eggs receiving transplanted nuclei began cleavage. When the *Xenopus* egg begins to develop, concurrently with the mitotic divisions the cytoplasm becomes divided up into first, two cells, then four, then eight, and so on. This is the process called *cleavage*. Nuclei were taken from the 25 per cent of partially cleft eggs for the second (serial) transfer to an egg identical to the first one (Fig. 4). Of an initial number of 129 nuclei taken from the keratinized skin cells, 6 lines gave rise to swimming tadpoles after the serial transfer, and all of the nuclei of the cells in these tadpoles had only one nucleolus.

cleavage

QUESTION Can you suggest what explanations other than totipotency Gurdon and his colleagues were trying to rule out when they used (a) keratin-producing cells and (b) a visible nuclear cytological difference between the nuclei of donor and host eggs?

ANSWER By taking their nuclei for transplantation from among a sample in which 99.9 per cent were clearly synthesizing keratin, they were making sure with a 99.9 per cent probability that the nucleus did come from a differentiated cell. They calculated that the probability that all the successful transplants came from the 0.1 per cent of nuclei not in clearly differentiated cells was about 1 in 10^{10}! Obviously, one cannot claim totipotency unless one is sure that the original nucleus came from a differentiated cell. Otherwise, it would still be possible to claim that the low success rate might reflect that not all nuclei were totipotent, making the successes a selected sample.

The nuclear 'marker', the difference in the number of nucleoli, was to ensure that the nuclei of the cleft eggs and of the tadpoles from the injected eggs really were characteristic of the skin cells and not simply a result of the failure to inactivate the 'host' nucleus in the injected eggs.

These quite sophisticated precautions strengthened the conclusion that the nuclei of a differentiated cell-type in *Xenopus laevis* do indeed remain totipotent. A 100 per cent success rate was not achieved. As Figure 4 shows, a number of first-transfer nuclei and serial-transfer nuclei did not lead to normal cleavage. One reason for this failure may be that the transplanted nuclei may not have stopped in the same phase of the cell cycle. When they begin mitotic divisions in the egg, some nuclei may be caught in the incorrect stage of the division cycle and some chromosomes or parts of chromosomes may not be replicated or may be left behind. This could give rise to aneuploidy (Unit 5, Section 5.2) in the embryo, which might be critical for development after the initial cleavages. Perhaps a certain chromosomal imbalance prevents successful development beyond the immediate cleavage stages.

QUESTION Can you suggest how we might test genetically the idea that defects in the nuclear genotype of an organism arrest development at a particular stage?

ANSWER There are at least three possible approaches that may have occurred to you. First, we might suggest that the chromosomal constitution of all transplants as embryos could be correlated with the extent to which they followed normal development, to see if successively more drastic chromosome abnormalities might be related to earlier deviations from normal development. Second, we might use nuclei from known aneuploid cells for transplantation to see how they fared. Third, we might use nuclei for transplantation that formed a gene product recognizably distinct from that of the maternal genotype (which produced the egg). We could then see when the gene product first appeared. If on a number of occasions with various genotypes, the difference failed to appear before a certain stage, we might be inclined to suggest that perhaps this was the first stage at which gene products from the transplanted nucleus were used. It would then not be surprising if failures in transplantation caused by inadequacies in the nucleus (e.g. aneuploidy) were not manifest until this stage.

Realize where this is leading. We have now accepted the experimental evidence on which totipotency is based. To explain the differences that emerge between different cell types (cellular differentiation) as well as the different morphogenetic processes, we are forced to invoke another hypothesis, that perhaps different genes are in action in different cells or at different periods during development. In the next Section (Section 8.3) we look at the way geneticists have tried to test this hypothesis. Before doing this, we should make a comment about the various organisms from which examples are taken in this Unit. *Xenopus laevis* and other amphibians, in particular species of *Rana*, have been used for nuclear transplantations because of the technical advantages of various aspects of their biology: their eggs are very large, unlike those of *D. melanogaster*, for example; fertilization and development is external, unlike the mouse, for instance, so that in a suitable medium the developmental changes can be followed and manipulated more easily. Nevertheless, nuclear transplantations have been made in *D. melanogaster* embryos despite their small size, and the inclusion of this organism increases the generality of nuclear totipotency. You will recall from Units 3 and 4, Sections 3.5 and 3.7, that different species were used to investigate various facets of meiosis, some being suitable for cytological examination and others useful because entire tetrads could be recovered together. The same is true in the analysis of genetic aspects of development; one species is excellent for one form of investigation, while being most intractable and unsuitable for another problem. It certainly is not contrariness that leads us to draw examples from different species!

8.2.2 Summary of Section 8.2

1 In a definitive experiment, nuclei from adult cells of the skin that were synthesizing keratin, and therefore differentiated, were transplanted to eggs. After serial transfer, these nuclei yielded normal functioning tadpoles. Because they are able to support the complete spectrum of development, these nuclei are said to be totipotent.

2 One of the precautions taken in the transplantation experiment was to use a cytological difference between the nuclei of the donor and host cells to eliminate cases of successful development due to the incomplete inactivation of the egg nucleus. This is a straightforward example of the application of a genetic method to clarify the experimental situation.

3 The suggestion was discussed that one possible reason for the failure of some transplanted nuclei was that, as a result of a transplantation 'shock', they had become aneuploid.

4 The corollary to the demonstration of nuclear totipotency is that cells can become different (differentiated) only through changes in the activity of their genes, as totipotency is a proof that the differentiated cell nucleus still contains a complete genome.

8.3 Variable gene expression as the basis for development

This Section contains evidence in support of the hypothesis that development is based on differential gene expression in time and space. In other words, different genes are expressed in different cells or in the same cell at different times in its history, and this activity results in the process we call development. You will be encouraged to criticize this evidence as we proceed. By gene expression, we mean some activity of the gene that can be measured and that can reveal allele substitutions when they are made. We exclude the actual *presence* of the gene in the cell nucleus as an activity. Other than this, one can include among the 'activities' the (possible) transcription of the gene to RNA, its (possible) translation to an amino-acid sequence and any subsequent modifications and activations of the product. 'Possible' appears in parentheses because not all genes (defined by functional and mapping tests) are necessarily transcribed or translated (neither tRNA nor ribosomal RNA are translated). As you will see later, many intensely studied systems still leave us a long way from any knowledge of the biochemistry or the molecular biology of gene expression. In such situations, defining the period when a gene is expressed may still leave us in the dark about *which* of the molecular activities—transcription, modulation, translocation of RNA, translation, activation or degradation of the mRNA—we have indirectly been measuring.

The evidence presented here is an answer to the following questions.

1   If we examine mutations that are lethal to the organism that carries them, do these mutations give any clues about the temporal activity of genes in development?

2   If we compare the activity of the same gene in different cells of the same individual, do we find any differences? Does the activity of particular genes change over a period of time in the same cell?

3   Does the cytological appearance of chromosomes at different times in the life span of the same cell give any clues about changes in gene activity?

8.3.1   Recessive lethal mutations

Lethal mutations have a clear-cut and unambiguous phenotype; they lead to the death of the organism carrying them. It turns out to be easy to induce and collect many lethal mutations in different genes. You remember that we know that two recessive lethals are in different genes if the heterozygote formed between them is not itself lethal.

What can we learn about gene action during development from a collection of lethal mutations? Let us consider some straightforward experiments that could be carried out with each mutation. Imagine that we cross male and female flies, all of which are heterozygous for the same recessive lethal, and that the other homologous chromosome is a balancer chromosome (Unit 5, Section 5.3), which itself contains another recessive lethal mutation.

> QUESTION   How many different genotypes survive to be fertile adults in such a mating?

> ANSWER   Only one genotype survives: the heterozygote identical to the parents. Both homozygotes formed are lethal and the balancer chromosome ensures that no recombination occurs between the two lethal mutations, one in the balancer and the other in the normal homologue.

From this balanced lethal stock, we construct genotypes that are heterozygous for the lethal and for a chromosome of normal gene sequence that carries no known lethals (see below). These flies are mated to each other and samples of eggs laid by the females are collected (see (a), below).

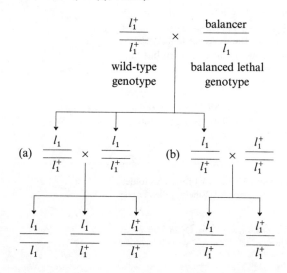

345

A second mating (b) is slightly different, being between a heterozygote for the lethal mutation and a wild-type genotype. Samples of 500 eggs are taken from both matings (a) and (b) and the number of eggs that successfully hatch into larvae is noted for each type of mating. For other samples of eggs, the number of adult flies that emerge after the larvae have pupated and metamorphosed is recorded. These data for matings (a) and (b) are shown in Table 1. (The life cycle* of *D. melanogaster* shows you how long these stages last.)

Table 1 Survival of progeny from matings (a) and (b)

Mating	Number of eggs laid	Number of eggs hatched	Number of adults
(a)	500	480	373
(b)	500	491	479

ITQ 2 Select the correct interpretation of the data in Table 1. The recessive lethal homozygote dies

(i)   after it emerges as an adult

(ii) during the embryonic period before hatching from the egg to the larva

(iii) after hatching from the egg but before emerging as an adult?

Use χ^2 tests (*STATS*†, Section ST.4) to judge whether the results from matings (a) and (b) are different.

This experiment has allowed us to define the *lethal phase* of the recessive lethal mutation, that is, the developmental period during which death ensues—although it is generally observed that abnormalities appear in the developing organism *before* death. By careful observation of individual putative homozygous lethal embryos as they develop, it is possible to define the first point at which their pattern of development deviates from that seen in wild-type embryos; this point is called the *phenocritical period*. It may precede the actual lethal phase by quite some time for some lethals. Figure 5 shows the phenocritical periods for a large number of different recessive lethals in *D. melanogaster*, which die as embryos.

lethal phase

phenocritical period

The extent of the shaded bar for each mutation indicates the period during which development is normal; the phenocritical period is the point at the far end of this line.

ITQ 3 Each of the statements below interprets the data given in Figure 5. With which of them do you agree?

(i) The normal (wild-type) activity of all these genes is required somewhere before the initiation of the corresponding phenocritical period.

(ii) The phenocritical period defined above indicates when the lethal gene is first transcribed and translated, and the data as a whole provide good evidence supporting the idea of phase-specific gene activity in development.

(iii) The phenocritical period is the developmental period when the result of using the defective gene becomes critical; it may not necessarily indicate the *first* use of that gene, but rather the complexity of the developmental process at that point. The defective gene product is 'shown up' at this point but may have been expressed continually from fertilization.

(iv) The phenocritical periods of different genes are distributed *throughout* the embryonic period and have the appearance of a collection of *random events* and thus cannot be evidence that organized differential gene activity is the basis for development.

* The Open University (1976) S299 LC *Life Cycles*, The Open University Press. This folder, containing details of organisms mentioned in the Course, is part of the supplementary material for the Course.

† The Open University (1976) S299 STATS *Statistics for Genetics*, The Open University Press. This text is to be studied in parallel with the Units of the Course. We refer to it by its code, *STATS*.

(v) These data can never be used in support of the hypothesis that differential gene activity is the basis of development because they are a selected sample of lethal mutations. Only embryonic lethals were included in the sample, so the prediction on the basis of this hypothesis is self-fulfilling.

So although this pattern of phenocritical periods of different embryonic lethal mutations is good circumstantial evidence for differential gene activity during embryogenesis in *D. melanogaster*, without being able to identify the appearance of the primary gene products involved, we cannot claim that this is definitive evidence. Even so, we would find very few developmental biologists who would not now accept our hypothesis as a very strong working basis for part of a theory of development.

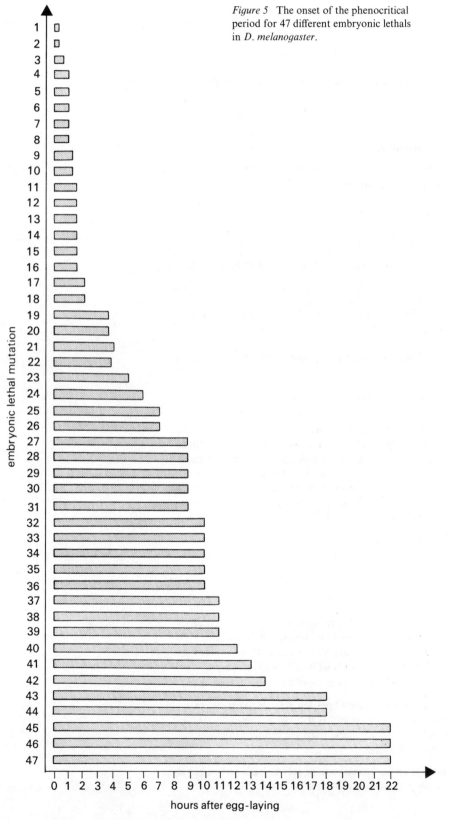

Figure 5 The onset of the phenocritical period for 47 different embryonic lethals in *D. melanogaster*.

Where numbers of lethal mutations have been examined (and there is similar, valuable, information from the house mouse *Mus musculis*, as well as from *D. melanogaster*), they substantiate the picture drawn about embryogenesis. However, in *D. melanogaster*, the distribution of phenocritical periods is not spread at all evenly throughout the $9\frac{1}{2}$ days of development from egg to adult. Clusters of phenocritical periods are found at the time of hatching from the egg, at pupation and at the time of hatching of the adult. Again, this observation has made geneticists cautious because perhaps these points should be seen as 'crisis periods' in development when defects that have existed for some time begin to tell.

Now, there is a way of refining the determination of the phenocritical period so as to make this point clearer and that is through the use of what are called 'conditionally lethal' mutations. It is possible to find mutations that are lethal only under particular environmental conditions, for example, at more elevated temperatures than are normal for the species. A *temperature-sensitive lethal mutation* (often abbreviated as *tsl*) manifests itself as a lethal at a higher (or non-permissive) temperature, but is not lethal at the normal (or permissive) temperature (see Table 2).

temperature-sensitive lethal mutation

Table 2   The contrast between the behaviour of lethal and temperature-sensitive lethal mutations in *D. melanogaster*

	Lethal mutation			Temperature-sensitive lethal mutation		
Mating	$\dfrac{\text{balancer}}{l_1}$ $\times$ $\dfrac{\text{balancer}}{l_1}$			$\dfrac{\text{balancer}}{tsl_1}$ $\times$ $\dfrac{\text{balancer}}{tsl_1}$		
Progeny	$\dfrac{\text{balancer}}{\text{balancer}}$	$\dfrac{\text{balancer}}{l_1}$	$\dfrac{l_1}{l_1}$	$\dfrac{\text{balancer}}{\text{balancer}}$	$\dfrac{\text{balancer}}{tsl_1}$	$\dfrac{tsl_1}{tsl_1}$
Fate at 25 °C (normal temperature)	dies	lives	dies	dies	lives	lives
Fate at 29 °C (non-permissive temperature)	dies	lives	dies	dies	lives	dies

The advantages of a conditional lethal mutation in developmental analysis are made very clear in TV programme 8. It is sufficient to say here that it is possible to define the period during development when growth in the non-permissive environment (e.g. at the elevated temperature for a *tsl* mutation) leads to the death of the homozygote *tsl*/ /*tsl*. All that is done is that different samples of the *tsl* homozygote are moved from the customary temperature of 25 °C, to 29 °C at different points in the developmental process. Interpretation of the results allows the geneticist to define the temperature-sensitive period for this gene product.

8.3.2   Puffing in polytene chromosomes of *D. melanogaster*

In Unit 5, Section 5.1 we introduced you to the structure of the polytene chromosomes found in certain cells of the larvae of various dipteran flies, including *D. melanogaster*. In some chromosome preparations, certain of the chromomeres, or 'bands', appear swollen or 'puffed up'. Figure 6 shows what is meant by a puff—an area of the chromosome stains more diffusely and appears larger in (a) than in (b).

Examination with the light and electron microscopes has revealed that the many parallel DNA strands appear to be unravelled and that there is granular material associated with the DNA strands in the puff. These puffs are believed to be sites of RNA synthesis on the chromosome because:

1   Pulse-labelling with tritiated uridine, a precursor of RNA, leads to the rapid appearance of intense radioactivity specifically at the puffed regions.

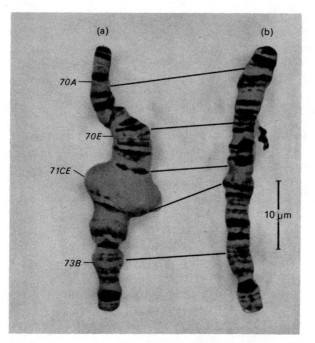

Figure 6 Photomicrograph (retouched) of part of a polytene chromosome from *D. melanogaster* showing chromomeres in (a) the puffed and (b) the unpuffed state.

2 Stains that are specific for RNA reveal the puffs, but after the chromosome has been exposed to RNase, the puffs no longer stain.

> QUESTION Do you consider this evidence, together with the information from Unit 5 on genes and chromomeres (Unit 5, Section 5.5) to be sufficient to say that puffs must be the sites of RNA transcription from specific genes?

> ANSWER Although entirely *consistent* with this idea, these experiments do do not *prove* that the RNA is actually transcribed at the puff. The RNA could, for instance, be moved to the puff immediately after synthesis elsewhere.

The establishment of two points would clinch this discussion:

1 There is a direct link between the nucleotide sequence of the DNA in each chromomere and the RNA 'transcript' that appears at the puff.

2 At least one of these RNA products actually contains a messenger RNA sequence.

The first step has been achieved, essentially in the following way. Newly synthesized RNA from a particular giant puff in the midge *Chironomus* has been separated by micro-isolation of the tip of the chromosome carrying this puff. This RNA has been made radioactive by previously supplying the cell with radioactive RNA precursor molecules. Another sample of the same chromosome is prepared on a slide using a treatment that separates the complementary strands of the DNA fibres. When the radioactively labelled RNA is supplied to this preparation under specific conditions favouring base-pairing, the RNA binds to the chromosome at any point where there is a sequence of bases complementary to itself. Subsequent autoradiography of this chromosome shows that the material is localized at the same puff from which the RNA was taken. Thus, it is extremely likely that the RNA was actually transcribed from that sequence and, more generally, we would conclude that puffing does indicate RNA transcription at that point.

This conclusion is very important. If the appearance of a puff is an indication of gene activity (specifically, transcriptional activity, of course) in that chromomere, then the examination of the distribution of puffs can be used to tell us which genes are active and at what time they are active. In other words, we are able to visualize genetic activity cytologically. Figure 7 shows a series of preparations of the same region of a polytene chromosome from the same genotype, but taken from different individuals, which were at different points in their development. The sequence represents the period of time from 110 hours after egg-laying until 12 hours after the formation of the prepupa.

The prepupa is an immobile stage that appears before the true pupa is formed.

The pictures in Figure 7 are really a developmental history of the puffing activity of that chromomere. It is possible to obtain similar information on the state of other chromomeres, and in Figure 8 the puffing activity of 23 chromomeres in chromosomes of *D. melanogaster* has been summarized.

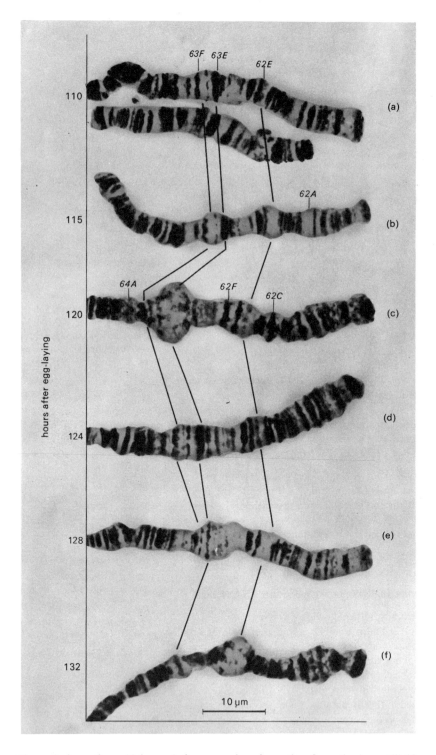

Figure 7 Photomicrographs (retouched) of the same region of chromosome III in *D. melanogaster* at different times, from 10 hours before the formation of the prepupa until 12 hours after the formation of the prepupa. Identical chromomeres are joined by lines.

Figure 8 shows from 10 hours before puparium formation from the larva till 12 hours after that event. The size of the puffs seen are expressed as the height of the blocks; a taller block is a larger puff. You will have noticed that there are changes in the size of individual puffs with age, but that although some may show an increase first and then fall off (*78D*), others puff twice in this period (*74EF*) and yet others remain more or less constant (*72CD*). Figure 8 represents only a small sample of the chromomeres that do show puffs, but of the total, about 80 per cent of the puffs specifically appear at a particular time in development. Another way of putting this is to say that the occurrence of puffing, taking the entire genome together, has a temporal pattern to it. The nuclei of the salivary-gland cells are not the only ones with polytene chromosomes in the larva—polytene chromosomes can be found in the fat body and in the malphigian tubules. Chromosomes in nuclei of those tissues also display puffs, but the pattern is in detail different from that in the salivary glands.

Now, have we lost sight of the wood for the trees in discussing this peculiar cytological phenomenon? Not at all! Let us recall the steps in the argument established so far.

1 Each chromomere seems to be a genetic functional unit.

2 Puffing of a chromomere is evidence that RNA transcription is occurring.

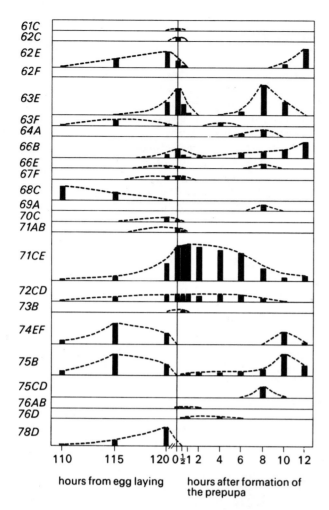

hours from egg laying

hours after formation of the prepupa

Figure 8 Diagrammatic representation of the puffing activity of chrommeres in *D. melanogaster*, from the end of larval life until the formation of the prepupa. The height of each block indicates the magnitude of the puff.

3 The pattern of this transcriptional activity varies between chromomeres with time and also relates to the particular tissue from which the chromosome comes.

It is fair to conclude from this that puffing activity is a reflection of specific genetic activity and that, in turn, this activity is varying with developmental age and according to the tissue-type from which the chromosomes are taken. This system in *D. melanogaster* offers a unique situation in which genetic, cytogenetic and biochemical tools can be turned together on to the question of gene expression. You may wonder why one should find evidence for transcriptional activity in cells that are part of a gland due to be broken down during metamorphosis in the pupa, and why the chromosomes are polytene when in the majority of other cell types in the larva the nuclei remain diploid. It has been suggested that the salivary glands might have a 'template' role during metamorphosis as sites at which extra transcription is required and at which the products or translated products are to be used by other tissues destined to be part of the adult.

8.3.3 Summary of Section 8.3

1 Examination of segregating progenies, including recessive lethals, permits the definition of the lethal phase (the time of death) and the phenocritical period (first visible abnormality) for any mutation.

2 The distribution of phenocritical periods for a collection of recessive lethals is consistent with varying activity of different genes at different times in development.

3 The analysis of conditionally lethal mutations under permissive and non-permissive conditions permits identification of the period during which that gene is critically important for the development of the organism.

4 Puffs in chromomeres of polytene chromosomes are visible evidence of transcriptional activity in those chromomeres.

5 There is a defined sequence and pattern to puffing activity in different chromomeres during development, which is entirely consistent with differential gene activity as the basis of development.

8.4 Genetic composites: mosaics and chimaeras

It is often noted that the more obvious changes in a mutant are localized to a particular tissue or region. For instance, the changes may apparently involve only the wings or the eyes in a fruit fly, or solely the leaf shape in a plant, or only the coat colour in a mouse. One of the basic concepts we are examining is that different genes may be active in different parts of a developing organism and we ask why those differences occur. However, the descriptive evidence from the sort of mutants just mentioned, showing changes that are restricted to part of the organism, is not sufficiently good to support the idea that the activity of the particular gene being scrutinized is, therefore, limited to that tissue. Consider the hypothetical case of a gene that contains the nucleic acid sequence that codes for a protein hormone produced in the brain of an organism. Let us further suppose that this hormone is intimately involved in the co-ordination of growth in the limbs. We could imagine a mutation in this gene resulting in a less efficient hormone and that this disfunction was most noticeable in the limbs, which were distorted in shape. On the face of things, such a mutation might easily be interpreted as a gene active in limb tissue, whereas we stated that it was actually transcribed and translated in brain tissue. In its underlying cause and explanation, this mutation is very different from a genetic defect intrinsic to limb tissue, which would prevent it from achieving normal cell division and growth. The distinction being made here may be plainer if we consider the example to apply to man. If one wished to forestall or remedy such a defective situation then it would be all important to know the focus of activity of the gene, whether it was in fact in nervous tissue in the brain or in the cells of the growing limb.

How does one localize the tissue or cells in which a gene has its effect? We shall call this localized area the *focus* of the mutation: the cells in which the expression of the mutation (by transcription and translation) results in the defect. One answer is to try to construct a *genetic composite* individual, one in which not all the cells are of the same genotype—the normal situation. A simple analogy makes clear how this would be useful. Imagine that your desk lamp suddenly fails to light up (it is a 'mutant' phenotype), and you are anxious to pin-point the fault (find the 'focus' of the mutation). One takes a second but working desk lamp (a 'wild type') and proceeds to exchange components ('tissues') between the two lamps (the plug, the bulb, the flex, and so on) producing a series of ('genetic') composite lamps. When such an exchange ('genetic composite') restores the defective lamp to working order, but prevents the previously normal lamp from working, then you would know you had localized the defect (the 'focus'). In a similar way, provided we can construct a genetic composite organism with some cells of normal genotype and other cells carrying the mutation and can identify the source of the cells in the composite, then we shall soon discover in which cells the genotype must be normal for the abnormality not to be expressed. In addition, if we could also arrange that the defective genotype appeared in particular cells at different times after the initiation of development, then we might get an idea of how early this defective gene was acting.

focus of a mutation

genetic composite

Now this general argument about the use of genetic composites may not be clear to you until you meet a specific example. First, though, it is helpful to explain the ways in which genetic composites may be formed. Such individuals do arise spontaneously, although this is very rare; however, there are a number of genetic 'tricks' that we can use to bring about their formation.

8.4.1 Methods of forming genetic composite individuals

We have adopted the term genetic composites to include two distinct classes, genetic mosaics and chimaeras. *Genetic mosaics* are individuals that contain cells of at least two distinct genotypes, all derived from a single fertilized egg. At present, they are almost entirely restricted to *D. melanogaster*, and can be produced by two distinctly different processes, induced *mitotic crossing over* and the loss of chromosomes. Figure 9 shows how crossing over during mitosis can lead to the formation of a genetic mosaic. Figure 9, stage 1 represents the two X-chromosomes in a diploid cell of a developing *D. melanogaster* embryo that is heterozygous for the recessive mutation 'yellow body' (y/y^+). Adults of this genotype will be phenotypically grey-bodied. In Figure 9, stage 2, this nucleus has undergone chromosome duplication before mitosis, but has, in addition, been subjected to a brief exposure to X-irradiation, which induces breaks in the chromatids. In dipteran insects like *D. melanogaster*, homologous chromosomes pair up in somatic cells. Remember that homologues do not normally do this at mitosis, but only at meiosis.

genetic mosaic

mitotic crossing over

If non-sister chromatids happen to break and rejoin with opposite partners, mitotic crossing over will have occured (Fig. 9, stage 3.). The two centromeres attached to sister chromatids move to opposite poles at anaphase of the next mitosis and because of the induced exchange event, this mitosis produces two different diploid daughter nuclei, one homozygous for the mutant allele $(y//y)$ and the other homozygous for the wild-type allele $(y^+//y^+)$ (Fig. 9, stage 4). This is the basis of the process and it is thought to depend upon the occurrence of somatic pairing and thus is restricted to diptera. But can such a genetic mosaic be recognized in the adult fly?

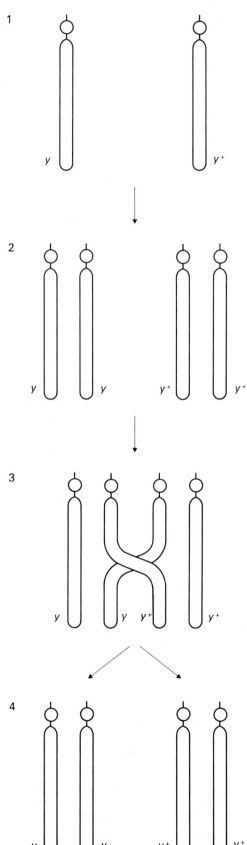

Figure 9 Formation of a genetic mosaic by mitotic crossing over during cell division in a heterozygous genotype of *D. melanogaster*: stage 1, diploid genotype heterozygous for yellow body (y); stage 2, chromosome duplication before mitosis; stage 3, chromatid breakage and exchange induced by X-rays, and the separation of chromosomes at anaphase; stage 4, two non-identical genetic products of mitosis.

Both the daughter cells (Fig. 9, stage 4) continue to divide by regular mitosis to form two *clones*. Both of these clones will constitute some part of the adult tissue; but it turns out that the cells of genotype $y//y$ form the yellow colour characteristic of this mutation even when they are in a mosaic with cells that will produce the grey body-colour ($y//y^+$) and the $y^+//y^+$ cells formed by the same mitotic crossing over. A gene is said to be *cell-autonomous* in its action if the two distinct phenotypes of the two genotypes are produced by the two types of cell when they are adjacent to each other. The gene yellow body is cell-autonomous in *D. melanogaster*, and you will read shortly how geneticists have capitalized upon this fact.

clone

cell-autonomous (gene action)

The other way in which a genetic mosaic can be formed is by the loss of one of two homologues in a heterozygous genotype during a nuclear division (Fig. 10).

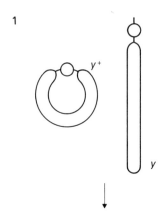

first mitotic division in the egg during which one
ring X – chromosome fails to be transmitted

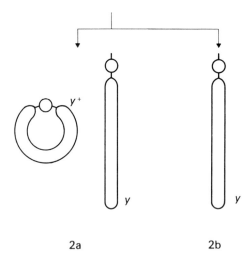

2a   2b

Figure 10 The formation of a genetic mosaic by the loss of the ring-shaped X-chromosome at the first division in the fertilized egg of *D. melanogaster*: stage 1, two X-chromosomes, one ring-shaped and one rod-shaped in the zygotic fusion nucleus; stage 2a, the XX daughter nucleus, which forms grey body-colour in skin; stage 2b, the XO daughter nucleus, which forms the yellow body-colour.

The yellow-body gene lies at the distal tip of the X-chromosome and in this example one X-chromosome carrying the y^+ allele exists in a ring form and the other carrying the y allele as the more familiar acrocentric rod X-chromosome. It is an observed feature that many ring-shaped X-chromosomes fail to divide and separate correctly at the first division of the zygotic nucleus in the fertilized egg, which results in two nuclei, one with both X-chromosomes and the other which has lost the ring-shaped chromosome (Fig. 10, stages 2a and 2b). After this division the ring-shaped X-chromosome behaves normally at mitosis and so the embryo and eventually the adult fly will be composed of an equal number of cells containing each type of nucleus as a genetic mosaic.

Genetic chimaeras are individuals containing cells of more than one genotype, but formed by the mixture of cells derived from more than one fertilized egg. The methods of production are conceptually simpler than those involved in forming mosaics, but may be technically quite as difficult to achieve. They consist of creating mixtures of cells or nuclei of different genotype by the injection of cells or nuclei from one developing zygote into another, or by bringing about the fusion of two developing embryos of different genotype.

genetic chimaera

8.4.2 The cell-autonomy and period of activity of a gene determined by mitotic crossing over in *D. melanogaster*

This Section describes first some features of the development of the adult structures in *D. melanogaster*, then a rather bizarre change produced by mutation and, finally, an experiment using genetic mosaics including this mutation, which gives us insights about the genetic control of this developmental process.

D. melanogaster is a dipteran insect and one of the taxonomic features is the possession of two wings. There are mutations that cause the fly to develop two extra wings in place of the halteres on the thorax. Figure 11 shows the appearance of such a fly, which is carrying mutations in a complex series of genes called the 'bithorax' complex in the third chromosome.

As well as having two pairs of wings, this fly has a duplication of part of the thorax from which the wings arise. The bithorax mutation and several others like it have long intrigued and baffled developmental biologists as their phenotype is an exact and complete transformation of what would be particular appendages into recognizably different ones, for example, legs instead of antennae, wing tissue instead of an eye. These mutations are called *homoeotic mutations*. To understand the possible interpretations of these mutations we have to know more about the development of the external features of the adult. Each of the parts of the adult's surface, the legs, wings, eyes, antennae, for example, grow by cell division and develop during larval life from a closed sac of cells called an *imaginal disc* (Fig. 12). The imaginal disc

imaginal disc

Figure 11 Photomicrograph of the thorax of *D. melanogaster* homozygous for the mutant alleles of the 'bithorax' gene complex, showing the complete duplication of the wings and thorax.

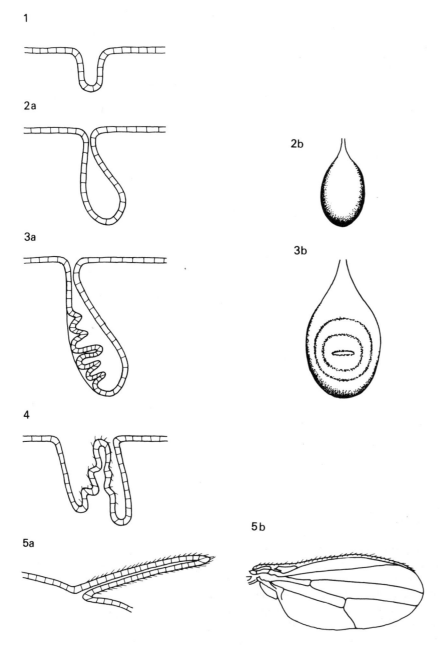

Figure 12 The formation and growth of the imaginal wing discs in *D. melanogaster*. Stages 1, 2a, 3a, 4 and 5a represent cross-sections through the disc showing its growth, folding and final eversion. Stages 2b, 3b and 5b indicate the shape and appearance of the disc and the wing derived from it.

itself arises as an infolding of the blastoderm surface during embryogenesis (Fig. 12, stage 1) and from that time onwards grows in virtual isolation until metamorphosis in the pupa when the single-cell layer of the disc folds inside-out and joins up with the products of the other discs (Fig. 12, stage 4) to constitute the surface layer of the adult (Fig. 12, stages 5a and 5b). These cells in turn secrete the rigid cuticle with its microscopic sculpturing, hairs and bristles.

There is long-standing evidence that the cells composing each imaginal disc (the eye-antennal discs, the six leg discs, the wing discs, and so on) are 'committed' or determined quite early on in embryogenesis to form one and only one such component of the adult. The evidence arises from the following procedure (Fig. 13).

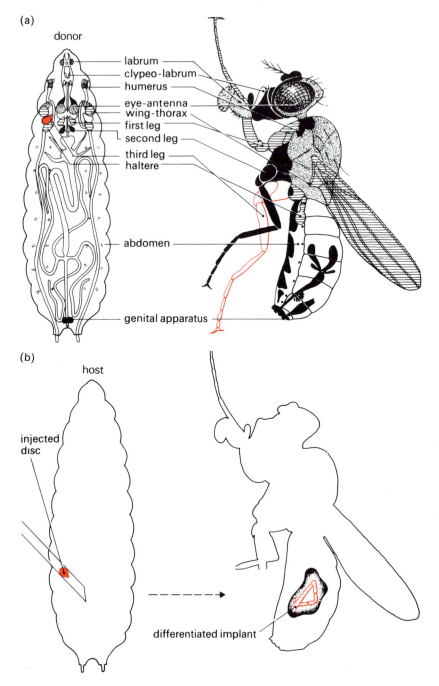

Figure 13 The steps involved in demonstrating that cells of imaginal discs in *D. melanogaster* are determined: (a) the location of the imaginal discs in a larva and how the various discs form particular structures in the adult; (b) the transplantation of one imaginal disc to a second larva and the formation of adult structures from the disc in the abdomen of the metamorphosed host.

If a larva is dissected, a set of imaginal discs can be found inside (Fig. 13(a)). There is a fixed number of discs of characteristic shape, and if a particular disc is removed from a dissected larva and implanted into a second larva using an injection needle (Fig. 13(b)), that disc develops in parallel with its host. When the host larva metamorphoses to an adult, the implanted disc can be found inside the abdomen of the host fly and has formed a complete set of adult structures characteristic of a particular area of the adult (a leg, an eye and antenna, or a wing). Each individual disc, characterized by its size, shape and location in the larva, always forms the same adult structure even though it has been removed from the influence of its 'sister' discs and donor larva by transplantation. It is this constant result that is the basis of the claim that each disc must already have been *determined*. Although the cells of all imaginal discs look similar under the microscope this transplantation test defines that they are determined for different adult structures. The relationship between individual imaginal discs and the structures they form is shown in Figure 13(a).

determined cell

Let us now see how homoeotic mutations like bithorax can be interpreted. They might be candidates for any of the following processes.

1 An error in the events in the embryo that initially give rise to the discs.

2 An error in the response of the discs to the hormonal 'triggers' involved in metamorphosis and transformation of the 'sac' of cells to the correct differentiated adult structure.

3 An error in the perpetuation of the determined state, whatever that might be in molecular terms, as the cells of the imaginal disc divide during larval life.

The mutation bithorax (*bx*) is recessive and it has been possible to construct a genetic mosaic larva so that part is homozygous for this gene (*bx*/ /*bx*) and part is heterozygous (*bx*/ /*bx*+). This enabled Morata as recently as 1973 to ask whether cells of *bx*/ /*bx* genotype can exert their peculiar homoeotic effect when surrounded by wild-type tissue (*bx*/ /*bx*+). The genetic mosaic was produced by inducing mitotic crossing over with X-irradiation. But it was crucially important to know when such a mosaic had been formed and so it was arranged that the *bx* allele was linked to other recessive alleles of genes that could be recognized on the body surface of the adult. Figure 14 shows how these mosaics could be recognized.

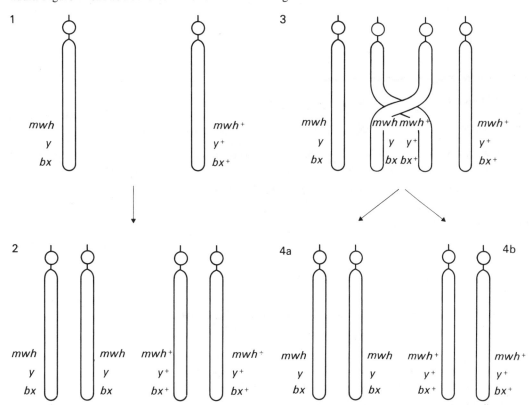

For this explanation certain features of the technique, in particular, the location and arrangement of the genes mentioned, have been simplified, but without loss of the essential detail of the procedure. The *bx* allele was distal to the recessive mutant alleles *y* (yellow body) and *mwh* ('multiple wing-hairs'—this mutation affects the hair-like processes formed on the wing surface) (Fig. 14, stage 1). Both these latter two mutants are known to be cell-autonomous in their effect, that is, in mosaics, patches of cells of genotype *y*/ /*y* and *mwh*/ /*mwh* can be recognized as distinctly different from the appearance of the wild type. X-irradiation produced some chromatid exchanges proximal to the *mwh* gene (Fig. 14, stage 2) and if the chromatids disjoined as shown in stage 3 of Figure 14, this resulted in two cells with new genotypes. Each of these cells grew and divided, as did the neighbouring cells, so that eventually the cell of genotype *bx y mwh*/ /*bx y mwh* formed a clone that was recognized as a patch on the surface of the adult. This patch was recognized because it was yellow and showed the multiple wing-hair effect.

Figure 14 The formation of cells of genotype *bx*/ /*bx* in an individual of genotype *bx*/ /*bx*+ by induced mitotic crossing over: stage 1, a homologous pair of chromosomes heterozygous for three genes; stage 2, chromosome duplication before mitosis; stage 3, X-ray-induced exchange between two chromatids. Stages 4a and 4b show two daughter nuclei of different genetic constitution.

In the experiment carried out by Morata, different batches of eggs or larvae were irradiated at different ages, so initiating clones that could be recognized in the adults eventually formed. Remember that the *bx* allele when homozygous transforms part of the haltere to a wing structure. The result is illustrated schematically in Figure 15, in which the appearance of mosaics involving the haltere is shown.

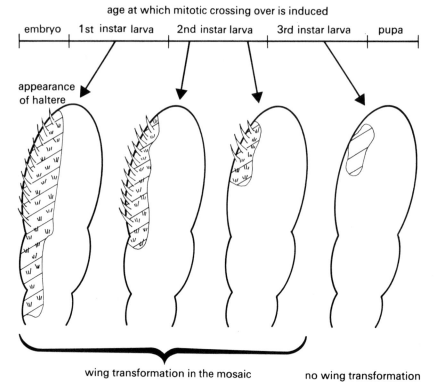

wing transformation in the mosaic   no wing transformation

a clone of cells showing the phenotypes of yellow and multiple wing-hair

wing bristles and appearance of multiple wing-hair phenotype in wing tissue

Figure 15 Schematic representation of the result of forming mosaics including *bx*//*bx* genotypic tissue in halteres of genotype *bx*//*bx*⁺ during larval development.

At almost any time in development that the mosaic was initiated, the clone of cells that was yellow and multiple wing-hair was also transformed, that is, it had the phenotypic characteristics of wing tissue and not of the haltere tissue surrounding it (Fig. 15). The only exception was that if the mosaic was initiated by X-irradiation less than about 24 hours before pupation then the transformation to wing was not seen, though the clone was still yellow and of multiple wing-hair phenotype. At 24 hours before pupation the cells of the discs have approximately three more cell divisions before they cease dividing.

Figure 16 shows an actual mosaic. The details of wing bristles and hairs formed in the haltere are shown in (a), and (b) is a lower magnification to indicate the relationship of this haltere to the thorax.

ITQ 4 Remember the question we posed several paragraphs ago about the time and site of action of the *bx* gene. Referring to Figure 15 for information, classify the following statements as true, untrue or possibly true.

(i) The fact that every clone of the marker phenotype is also transformed to wing, save those induced very late, indicates that the bithorax gene is cell-autonomous in its expression.

(ii) Because even *bx*//*bx* clones formed *late* in development, except during the last few cell divisions, still show the transformation, this means that the mutation is more likely to be an intrinsic and autonomous change in the response of the disc tissue during metamorphosis.

(iii) The fact that a mosaic produced *early* in development shows wing tissue proves that the gene is expressed *early* and is, therefore, likely to be affecting the initial determination event in the embryo.

(iv) Mosaics produced at the very end of larval development show the yellow body and multiple wing-hair phenotype, but not the transformation to wing

358

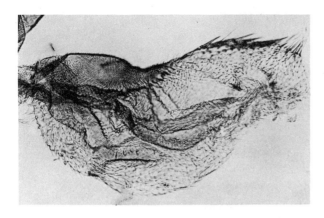

because the ensuing two or three cell divisions are not sufficient to 'dilute out' the wild-type gene products of the bithorax gene sufficiently to show up the mutant phenotype characteristic of $bx/\,/bx$.

(v) Mosaics produced at the very end of larval development show the yellow body and multiple wing-hair phenotypes but not the transformation to wing because the bx gene is no longer expressed or used after this point.

Quite a considerable amount of information about the action of the bithorax gene has been discovered using genetic mosaics. It is important to remember that the ultimate aim is to explain the role of the bx^+ allele in development, although we investigate this indirectly by studying the effect of the mutation (bx). It is possible to rephrase the conclusion by saying that the role of the bx^+ gene appears to be to maintain the determination state of haltere cells during growth in the haltere disc. If the gene fails, then the tissue determination reverts to 'wing' type.

8.4.3 The use of genetic mosaics formed by chromosome loss in *D. melanogaster*

As part of the required reading for this Unit you will have received a copy of a paper by S. Benzer entitled 'Genetic dissection of behaviour', which you should read in conjunction with this Section. But first there are some points to note that will assist your comprehension. In this paper, Benzer is interested in analysing behaviour using genetic tools, but as we hope you will realize, it is difficult to draw a demarcation line between development and behaviour, and the main lessons to be learned are as relevant to the study of development as they are to behaviour. Do not spend time mastering the detail of the nervous system of *D. melanogaster*. There is also a section of the paper, from pp. 32–4, that is conceptually difficult and it is not necessary to understand the detail. You will probably find that the life cycle of *D. melanogaster* (in your *Life Cycles* folder) is helpful.

Now read the paper by S. Benzer.

The crucial aspect of this paper is the generation and interpretation of the genetic mosaic *gynander* flies—flies that are half-and-half male and female in phenotypic appearance and reflect their genetic mosaic composition. In the same way that yellow body and multiple wing-hair were used as 'markers' to indicate the presence and limits of genotypically different tissue in the experiment in Section 8.4.2, yellow is used in this paper. The formation of the mosaic is depicted in Figure 10 of this Unit, on p. 354.

gynander

> **ITQ 5** Read the statements, (a), (b) and (c) below, and for each statement select the correct explanation.
>
> (a) In the gynanders, the yellow-body phenotype indicates the presence of 'male' tissue and the grey (wild-type) marks the 'female' tissue.
>
> *Explanations*
>
> (i) This is explained by the cell-autonomous expression of the X-linked recessive allele (y) of the yellow-body gene.
>
> (ii) The Y-chromosome causes the differentiation of male characteristics in this species, and its loss leads to the formation of female characteristics in half the cells.

Figure 16 (a) Photomicrograph of a genetic mosaic in a haltere of *D. melanogaster* showing a large sector of 'wing' tissue in the haltere. (b) A lower magnification of the same preparation to show the relationship between the haltere and the thorax.

(b) There is an obvious demarcation line separating the two halves of the fly that show male and female characteristics, respectively.

Explanations

(i) Each and every mitotic division in the zygote forms one XX and one XO cell, when one X is a ring-shaped X-chromosome.

(ii) The two populations of cells of karyotypes XX and XO, formed as a result of the unusual first mitotic division, do not subsequently intermingle during development.

(iii) Cells of XX or XO karyotype form in a 'pepper-and-salt' fashion during cell divisions and eventually sort out to make two distinct areas, each of which is entirely XX or XO.

(c) The yellow-body and the male phenotypic characteristics of the surface of the gynander fly are used to indicate which cells will possibly be able to show the phenotypic effects of the developmental or behavioural mutation being studied.

Explanations

(i) The mutation being studied is chosen because it manifests itself only in males.

(ii) The mutation being analysed is dominant in its expression.

(iii) The mutation being studied is X-linked and in the same homologue as the recessive allele (y), but the ring-shaped X-chromosome carries the y^+ allele and the dominant wild-type allele of the mutation being analysed.

The paper by Benzer illustrates several points in addition to the central one of the application of genetic mosaics in the determination of the 'focus' of activity of a gene. First, it gives you an idea of the complexity and diversity of the mutations that can be found in this species, which offer attractive starting material for developmental and behavioural investigations. Second, the 'exploded' diagrams of the body-parts of the adult fly superimposed on the egg's surface indicate the area of the egg's surface from which those structures originally derive. Such a representation is called a *fate map*. This deduction is made solely on information on the position of the borderline dividing the adult into its mosaic parts as seen in a large number of gynander flies. This information was used to construct the fate map of the egg, the detail of which is shown on p. 33 in Benzer's paper. Though you are not expected to study the details of the procedure, fate-mapping is nevertheless an elegant and informative use of a biological curiosity, the formation of gynanders.

fate map

8.4.4 The use of chimaeric mice in developmental analysis

Induced genetic chimaeras (Section 8.4.1) are currently being used in the mouse, *Mus musculus*, to analyse aspects of development in a way similar to that described in *D. melanogaster*. It is possible to form chimaeras by mixing together two embryos at an early stage in their development, at a time when the embryos each comprise only four or eight cells. Alternatively, it is technically possible to inject a single cell from an embryo of one genotype into a second embryo of a different genotype. These methods are being applied to the following issues.

1 Genotypic differences affecting pigmentation are used to reconstruct the likely origin and clonal ancestry of cells that give rise to the coat of the mouse. For example, cells of an embryo homozygous for the 'albino' allele ($c//c$) (which if they developed alone would give mice with white coats) can be fused with an embryo that has the wild-type alleles for this gene ($C//C$). The resulting embryos develop into mice that include some in which there is clear striping on the coat, reflecting the underlying chimaera. The difference affecting coat colour is in fact being used as a 'tag' or marker with which to follow the fate of the various cells in the chimaera.

2 In a similar way, genotypic differences affecting a biochemical or cytological character that can be distinguished in individual cells within a tissue mass, can be used to reconstruct the origins of the internal organs. For example, suppose we form a chimaera by injecting a single cell of genotype $a//a$ that lacked a non-essential enzyme function, into an embryo of genotype $A//A$ that had the enzyme function. If a method existed for distinguishing $A//A$ cells from $a//a$ cells by the absence of the enzyme activity in the latter, then we would be able to discover to what parts of the organism the single $a//a$ cell had contributed its mitotic products.

The method has considerable potential for advancing our knowledge of the clonal ancestry and growth of particular tissues and organs in a species rather more similar to man than is *D. melanogaster* (and thus of more immediate medical interest). One of the problems that should not be overlooked is that in contrast to *D. melanogaster* where cells that are part of the same clone usually stay together and in which there is limited cell movement and shuffling, in mammals such as the mouse there may be considerable importance attached to cell migration or selective 'aggregation' of particular types of cell so that a clone of cells may not remain as a contiguous cell mass.

8.4.5 Conclusions and summary of Section 8.4

There is one point that cannot be over-emphasized when discussing the use of genetic composites. This is that prime importance is attached to the idea that particular phenotypic differences (e.g. fur colour in the mouse, or body colour in *D. melano-gaster*) are used as indicators of the presence of genetic mosaics. This can be true only if the gene difference being used as the 'marker' is cell-autonomous in its expression. If cells of different genotype in close association in a tissue mosaic can influence each other's phenotypes for the 'marker' character, then they are of no use as genetic markers. Once genes that are 'good' markers of the genetic difference have been found, they can be used to investigate whether other genes are also cell-autonomous in their effects.

This Section may be summarized as follows:

1 Genetic mosaics and chimaeras (together known as genetic composites) are individuals composed of cells of more than one genotype. Mosaics are derived from the cells of one embryo and chimaeras are formed from nuclei or cells of more than one embryo.

2 Genetic mosaics can be produced by induced chromosome loss or by induced mitotic crossing over. Genetic chimaeras are formed by cell mixing, or by injection of a nucleus or cell into an embryo.

3 Genetic composites can be used to discover the focus of a mutation—the particular tissue in which the mutation exerts its primary effect.

4 Genetic composites are used to define the time of action and the cell autonomy of particular genes.

5 Genetic mosaics formed by X-chromosome loss in *D. melanogaster* have been employed to define the prospective fates of cells from various locations on the surface of the early embryo (fate-map construction).

6 Chimaeric mice are used to reconstruct the clonal history of the development of particular organs and tissues.

8.5 Nucleo-cytoplasmic interactions

From what we know already, can we say anything about the role of the cytoplasm in development, in particular, about the role of egg cytoplasm? First, we know (Section 8.2) that the transplantation of a nucleus from a differentiated cell to an egg in *X. laevis* alters its behaviour most strikingly. If left *in situ* in the adult frog, the 'leg' skin cell patently does not form another frog, but after transfer to an egg, a nucleus taken from that cell gives rise not to a clone of skin epithelial cells, but eventually to a swimming tadpole. The same nucleus responds differently to different cytoplasms, and somehow during development identical nuclei formed by repeated regular mitoses give rise to a diverse collection of differentiated cells. We cannot escape the conclusion that, after initially being in a common cytoplasm, the identical genomes of each differentiated cell type somehow have found themselves in a different cytoplasm. We should, therefore, look to the cytoplasm, and the interaction between nucleus and cytoplasm, for the basis of the control of gene expression.

We have chosen two examples to illustrate the interaction between nucleus and cytoplasm, both of which involve the cytoplasm of the egg cell.

8.5.1 Protection of the germ-line in *Wachtiella persicariae*

Wachtiella persicariae is an insect, a gall midge, that has 38 chromosomes in its zygotic fusion nucleus. In the somatic cells of the adult female of this species there are regularly 8 chromosomes of the 38, and in males fewer still are found, yet the cells of the germ-line continue to have 38 chromosomes.

> **ITQ 6** Do you think that these facts are in agreement with the concept of totipotency of somatic cell nuclei, that we derived from the experiments with amphibia (Section 8.2)?

Our main interest is in seeing how this species achieves this change in chromosome number in its somatic cells in each generation (see stages 1–5 in Fig. 17).

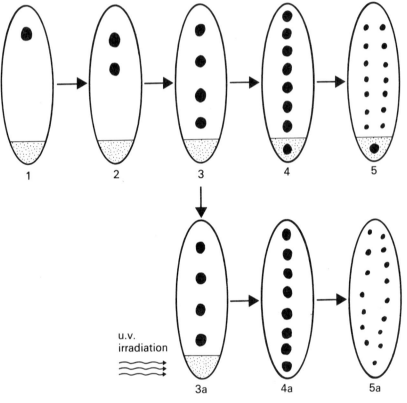

Figure 17 The consequences of ultraviolet irradiation of the poleplasm in *Wachtiella*. The normal sequence is shown by stages 1–5; large nuclei have 38 chromosomes and small nuclei contain 8 chromosomes. Stage 3a shows ultraviolet irradiation of the poleplasm; stages 4a and 5a show subsequent nuclear divisions.

In a similar fashion to *D. melanogaster*, the early cleavage divisions of the zygote nucleus are not accompanied by cell division, and the early mitotic products lie in a common cytoplasm. At the end of the third division (Fig. 17, stage 4) there are 8 nuclei each with 38 chromosomes, but one of these eventually finds itself in the *poleplasm* at one tip of the egg, a recognizably distinct area of cytoplasm. At the fourth mitotic division, all the nuclei outside the poleplasm undergo the elimination of chromosomes with the effect that the 14 daughter nuclei receive only 8 chromosomes each. These nuclei proceed to form the somatic cells of the midge. The nucleus in the poleplasm does not lose chromosomes and eventually forms the germ-line, the ovary in this case. What determines that one of those eight cells behaves differently? The distinct poleplasm holds the clue.

poleplasm

Geyer-Duzsinska performed an experiment to test the hypothesis that the poleplasm perhaps 'protected' the germ-line nuclei. She took eggs after the second cleavage division and irradiated the tip of the egg, including the poleplasm, with ultraviolet light, being careful to avoid exposing nuclei to ultraviolet (Fig. 17, stage 3a). This operation subsequently led to the elimination of chromosomes from all 16 nuclei at the fourth division (Fig. 17, stage 5a). She concluded that some factor in the cytoplasm protected the germ-line nuclei. Subsequently, it was possible to move the poleplasm by centrifugation and it was found that it always protected the nucleus that entered it.

We conclude that a distinct part of this egg cytoplasm has an important role in determining nuclear behaviour. This is a cytoplasmic difference that exists within a single cell, too. From what you will learn next, it is very likely that this heterogeneity in the egg cytoplasm was established when the egg was first formed in the ovary.

8.5.2 Genetic maternal effects

In our investigation of the role of genes in development we have reached a stage at which we are looking to the cytoplasm for the specific and different effects it can exert on the nuclear genome. The cellular continuity of the genome is beyond doubt but whence come the properties of the cytoplasm? Are we to look for self-perpetuating cytoplasmic entities comparable in stability with nuclear genes, or do the differences between various cytoplasms themselves come from previous differences in gene activity? A good place to start this enquiry is the egg stage, and we might rephrase the question in the following way: does the embryo develop from a genome sitting in a 'neutral' cytoplasm (a 'blank slate' as it were) or is there more to the egg than a physical container of the genome?

Consider this example, again taken from *D. melanogaster*. Below, we show three matings, (a), (b) and (c) between stocks with X-chromosome differences. *fs* is the symbol for a single gene mutation on the X-chromosome.

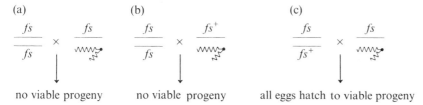

(a)   (b)   (c)

no viable progeny   no viable progeny   all eggs hatch to viable progeny

> **ITQ 7** What are the expected genotypes of the progeny for each of the crosses (a)–(c)? Select the correct interpretation of these three crosses from (i)–(v) below.
>
> (i) The *fs* gene has a dominant phenotypic effect and all progeny receiving at least one copy of *fs* die before becoming adults.
>
> (ii) When the female parent is *fs*/ /*fs* then no eggs develop no matter what genotype the eggs are.
>
> (iii) *fs* is a recessive lethal mutation.
>
> (iv) When the female parent is *fs*/ /*fs* only the eggs fertilized by sperm bearing *fs* fail to develop.
>
> (v) All eggs of genotype *fs*/ /*fs* fail to develop no matter what genotype their parents are.

The gene *fs* (short for female–sterile) represents a large class of recessive mutations in *D. melanogaster*, which when homozygous in females affect the viability of the eggs after fertilization, whatever their genotype.

Those effects are called *genetic maternal effects* because the viability of the egg depends upon the genotype of the female that laid it. It may help you to think of the egg (before fertilization) as part of the phenotype of the female parent, as it is strictly. It is a highly specialized cell-type! In which case it is not surprising to find genetic differences affecting the process of egg formation, or *oogenesis* as it is called. We have an immediate answer to our question about the role of the egg. It is obviously not just a package for the genome because there are important components supplied to it by the ovary, the absence of which would lead to the failure of development. Note that in cross (b) the introduction of a wild-type copy of the *fs* gene from the sperm did not remedy whatever defect is caused. This argues against the idea that the failure of egg development is the result of the lack of some common metabolite which could be furnished by gene expression from the wild-type allele entering the egg from the sperm.

genetic maternal effect

oogenesis

On the other hand, some genetic female–sterile maternal effects can be remedied by the introduction of wild-type alleles of genes from the male. The X-linked recessive mutation 'rudimentary wing' (*r*), which we mentioned in Section 8.1 and shall consider again in Section 8.6, is such an example.

An even more extraordinary genetic maternal effect is the mutation bicaudal in *D. melanogaster*, which you heard about in Section 8.1 (Fig. 3 on p. 341). Females homozygous for a particular mutation on chromosome II mated to any male produce a small proportion of eggs that form abnormal embryos having two sets of abdominal structures arranged end-to-end in mirror-image symmetry. There are no signs of head or thoracic structures at all.

8.5.3 Cytoplasmic maternal effects and intra-uterine effects

Two further maternal effects deserve to be mentioned here. You will be familiar with the general fact that mammalian embryos develop in the uterus of the female parent. The female parent supplies the metabolic needs and removes excreted products from the developing embryo via the placenta. A consequence of this 'support' system is that the embryo may be affected by changes in the ability of the maternal parent to sustain this constant internal environment, and we can here, in principle, distinguish two types of maternal effects. In Unit 7 you met an example of a cytoplasmic maternal effect—in the snail *Limnea* the direction of coiling of the snail shell was shown to be under the control of the maternal genotype (Section 7.6).

In mammals, it is possible to distinguish such a strict maternal effect on the egg structure from a change caused by a general alteration in the intra-uterine environment, the environment created and maintained by the exchange of materials via the placenta. Suppose we have discovered a maternal phenotypic effect in a mammal such as the mouse that means that all young born to mothers of a particular genotype ($a//a$) are different from normal young born to $a^+//a^+$ mothers. It is possible to transfer embryos at early stages between mothers of genotype $a//a$ and normal mothers $a^+//a^+$. Figure 18 shows this very schematically.

(a)　　　　　　　　　　　　　　　　　　　　　　　(b)

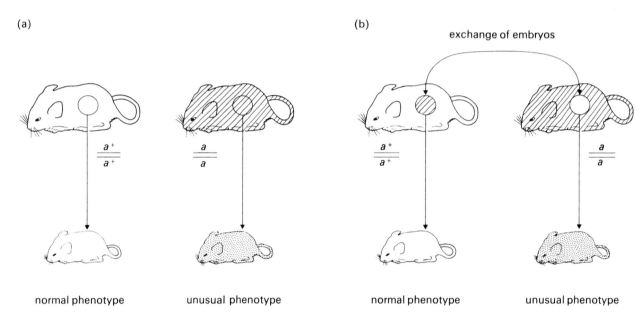

In this Figure, (a) indicates the common situation: a normal young mouse is born to a normal mother and a mother of a different genotype ($a//a$) gives birth to young with some particular phenotypic change (dotted). This is an effect of the maternal genotype, as it is not seen when the reciprocal cross is made and the $a//a$ genotype is in the male parent. If we take young embryos from the mothers shown in (a) and transfer them so that each embryo develops in the other maternal genotypic environment (see Fig. 18(b)), then we can distinguish a true cytoplasmic egg effect from an *intra-uterine environmental effect*, one due to the influence of a changed intra-uterine situation caused by the maternal parent.

Figure 18 Schematic representation of the result of the transfer of embryos from one genotype to another in *M. musculus*: (a) a normal mouse born to a normal mother, and an offspring of unusual phenotype born to a mouse of different genotype; (b) transfer of young embryos between the two genotypes during development.

> QUESTION Considering the result shown in Figure 18(b) offspring, do you think that this is due to a cytoplasmic egg effect or to the intra-uterine environment?
>
> ANSWER This result is an effect of the intra-uterine environment because the normal embryo $a^+//a^+$ developed the abnormal feature (dotted) characteristic of this maternal effect. The reciprocal transfer led to the formation of a normal offspring, even though of genotype $a//a$.

We have now come full circle in our argument in this Unit, having demonstrated that:

1 Phenotypic differences between cells during development are the result of differences in gene expression (Section 8.2).

2 There is evidence for differential activity of various genes at different times in development (Section 8.3).

3   Gene expression is conditioned by the cytoplasmic environment in which the nuclei lie (Section 8.1).

4   The cytoplasmic differences themselves may arise as the result of earlier gene activity (Section 8.4).

You may already have realized that this account forms the basis of a cyclical system of perpetuation of developmental processes. With the examination of mutations that have maternal effects we have opened the door on the investigation of the organization of the egg. Far from being a 'clean slate' on which the zygotic genome represents the 'writing', you may instead be beginning to see the egg as the apogee of the developmental process in the female of the previous generation. After all, as Samuel Butler, the nineteenth century biologist said, 'A hen is just an egg's way of making another egg'!

You will doubtless be curious about the molecular basis of these cytoplasmic effects upon the nuclear genome. We shall not give details in this Course but you may be interested to realize that the combination of the experimental technique of nuclear transplantation with the use of radioactively labelled 'host' eggs does offer an opportunity for identifying the molecules that pass into or out of the transplanted nucleus concomitantly with its change in activity.

It is worth pointing out that more than one genetic technique has been mentioned in this Section, although perhaps implicitly. First, there is the search and recovery of female–sterile mutations. Then, to establish that these were acting through the maternal genome, it was necessary to be able to manipulate the genotype of the maternal parent and of the zygote by genetic crosses to exclude other possible interpretations.

8.5.4   Summary of Section 8.5

1   Pole cells in insects are the cells that form the germ-line, the testis or ovary.

2   In *Wachtiella persicariae*, nuclei entering a specific region of the cytoplasm, the poleplasm, are protected from chromosome loss, which occurs in the formation of somatic cells in this species. The poleplasm may be said to determine the fate of the nuclei entering it.

3   Female–sterile mutations are a class of mutations that are expressed only in females during oogenesis, the formation of eggs in the ovary. These mutations result in the failure of the eggs to undergo normal development. They can be distinguished from normal recessive effects in the zygotic nucleus because they occur only if the particular genotype is present in the female parent.

4   Some genetic female–sterile effects can be removed or relieved if the sperm brings a normal wild-type copy of the gene in question into the zygotic nucleus, but other female–sterile mutations cannot be 'rescued' by any genotypic composition of the zygotic genotype.

5   In mammals, in which early development occurs in the uterus supported by the maternal parent, another maternal effect can be distinguished from a maternal cytoplasmic one, and this is due to failure to maintain a suitable intra-uterine environment (an intra-uterine environmental effect).

6   Intra-uterine environmental effects can be distinguished from strictly maternal cytoplasmic effects by examining the results of the transfer of embryos between different maternal genotypes.

8.6   Integration and interaction

A very important aspect of development has so far hardly had a mention. The process of development is integrated, that is, cells in particular tissues or structures can be shown to exert effects upon other cells in the developing organism resulting in the co-ordination of the developmental programme. This Section contains a number of examples of the interrelationship of 'parts' during development, chosen to show how genetic analysis can resolve what may be thought to be considerable complexity in the process of development.

8.6.1 Pleiotropy: effects of the mottled gene in *Mus musculus*

Figure 19 is of a male mouse showing the phenotypic effects of a mutation 'mottled-brindled' (Mo^{br}) which it carries on its X-chromosome. This male is a rare survivor of this genotype, the phenotypic effects of which are manifold and very severe. The hair pigment of the coat of this mouse is very diluted, although at the tips the hairs are pigmented, hence the name 'mottled' for this mutation. What is not obvious from this photograph is that, in addition, the structure of the hair is changed and the whiskers are curled. Such mice have a neurological disturbance; they have a sustained mild tremor and are rather inactive. Very severe phenotypic effects may also include skeletal defects, and males with some alleles of this gene do not survive for long after birth. This mouse is an example of the phenomenon of *pleiotropy*, a diverse and apparently unrelated collection of phenotypic effects, all caused by a single mutation. The separate abnormalities mentioned are called pleiotropic effects of the mutation.

pleiotropy

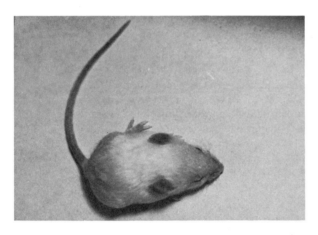

Figure 19 Photomicrograph of a mouse carrying the mutation mottled.

If you were trying to get to the basis of the mutation to work out the primary effect of this gene, where would you start? Quite a problem! Considerable genetic 'detective' work has been carried out on these mice and recently a common unifying and simplifying explanation has been put forward which makes sense of all the 'apparently' unrelated pleiotropic effects. The story runs as follows: Noradrenaline is a neuro-transmitter substance. It is released at nerve-endings and triggers electrical excitation in the next nerve cell across the gap between the cell membranes. In Mo^{br} male mice there is a severe deficiency of noradrenaline in the brain. This has been explained by a much reduced rate of conversion of the precursor molecule of noradrenaline, dopamine, to noradrenaline by the enzyme dopamine-betahydroxylase (DBH). Very curiously, comparison of the *in vitro* assays of DBH enzyme activity from Mo^{br} and normal mice showed that Mo^{br} individuals had an elevated activity of DBH! Careful examination of the biochemical method used to assay the enzyme activity indicated that the enzyme deficiency might be the result of a deficiency in copper in the brain tissue of Mo^{br} mice, which was present in the *in vitro* assay mixture.

> QUESTION Do you regard the demonstration of a copper deficiency affecting the enzyme DBH as a sufficient explanation for all the pleiotropic effects of Mo^{br}?

> ANSWER On the evidence we have presented, this is not a sufficient explanation because it is not clear how a neuro-transmitter defect would affect coat colour or skeletal growth.

It turned out that though copper is absorbed normally through the intestine wall in mottled mice, the concentration of copper carried in the blood-plasma proteins was reduced. Then a careful investigation of the role of copper in enzymic processes other than DBH showed that all the pleiotropic effects of mottled could be explained if Mo^{br} was a genetic defect in the transport of copper from the intestinal cells through the circulation.

The comments made in the introduction to Section 8.6 may now be clearer. Comprehension of the common physiological basis of all the pleiotropic effects of mottled mutations provides us with another picture of the interrelations (summarized; p. 367) of what might be thought of as independent developmental processes (such as hair growth and the development and function of the central nervous system). The interrelationships have been revealed by the analysis of the activity of one particular gene. Developmental genticists have not always been so fortunate in unravelling pleiotropic mutations (of which there are very many, particularly well documented in *D. melanogaster* and man as well as in *M. musculus*); as you may imagine, it is equally easy to choose to pursue a pleiotropic effect far removed from the primary cause and, therefore, that much more difficult to analyse!

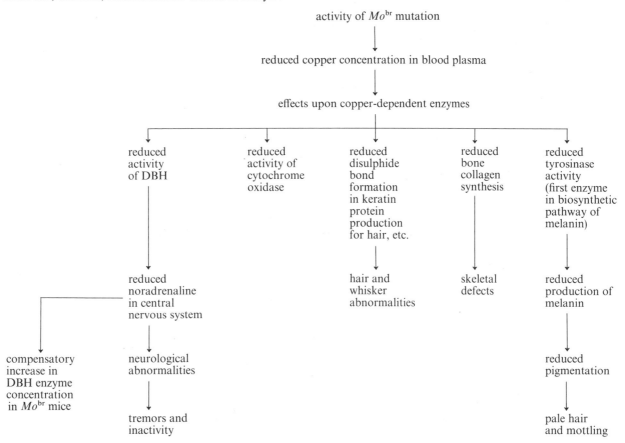

activity of Mo^{br} mutation

↓

reduced copper concentration in blood plasma

↓

effects upon copper-dependent enzymes

| reduced activity of DBH | reduced activity of cytochrome oxidase | reduced disulphide bond formation in keratin protein production for hair, etc. | reduced bone collagen synthesis | reduced tyrosinase activity (first enzyme in biosynthetic pathway of melanin) |

reduced noradrenaline in central nervous system

hair and whisker abnormalities

skeletal defects

reduced production of melanin

compensatory increase in DBH enzyme concentration in Mo^{br} mice

neurological abnormalities

reduced pigmentation

tremors and inactivity

pale hair and mottling

The mottled mutation in *M. musculus* appears to be very similar to a progressive brain disease in man, called Menkes kinky-hair disease, shown in Figure 20.

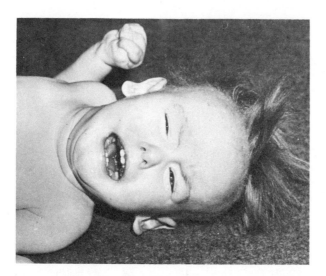

Figure 20 Photomicrograph of an individual suffering with Menkes kinky-hair disease.

The other phenotypic effects of this 'genetic disease' in man include abnormal white hair, changed hair structure and defects in the circulation of the blood. The point of mentioning this example of a parallel mutation in man is to emphasize the advantages to medicine if the general biology and causation of the defect can be investigated in a convenient laboratory organism like the mouse, in which the mutation can be perpetuated conveniently in breeding stocks. It is a small but interesting point that the mutations in both the mouse and man are X-linked, suggesting that the X-chromosomes in different mammals are perhaps homologous and carry similar genes.

8.6.2 Metabolism and development

What is the connection between the basic metabolic processes in cells and the process of development? Changes in genetic activity (the transcription and translation of genes) obviously must produce molecular changes in the cells involved. The interesting question to examine is how the molecular changes accompanying gene expression during development are related to the normal biochemical processes common to all cells. This idea was implicit in the preceding analysis of mottled. The difficulty lies in the fact that many of the genetic changes that affect development are not yet explicable in terms of molecular changes. For example, at the moment it is not possible to specify the molecular changes that accompany the homoeotic change, the development of a wing in place of a haltere in *D. melanogaster*. However, connections between a specific biochemical function and a developmental process have been made. The mutation Mo^{br} in *M. musculus* is just such an example, and another clear instance involves the mutation rudimentary wing in *D. melanogaster*, mentioned briefly in Section 8.1 and shown in Figure 2 on p. 340. During an investigation to see whether the nutrition of the female adults ($r//r$) was defective and so was causing failures in oogenesis, the mutant females were fed on various media. A particular baby food preparation cured the sterility of the eggs these females laid! It turned out to be the RNA in the food that was important and more refined tests showed that pyrimidines specifically restored fertility. This discovery naturally led to an examination of the activity of the enzymes involved in the biosynthesis of pyrimidines in $r//r$ females. The clear answer is now that activity of the first three enzymes of the synthetic pathway to pyrimidines in *D. melanogaster* is absent in various rudimentary mutants. Not all r alleles lack all three enzymes, but the general relationship between the genetic fine-structure of the rudimentary-wing gene and the enzyme deficiencies is set out in Figure 21.

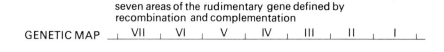

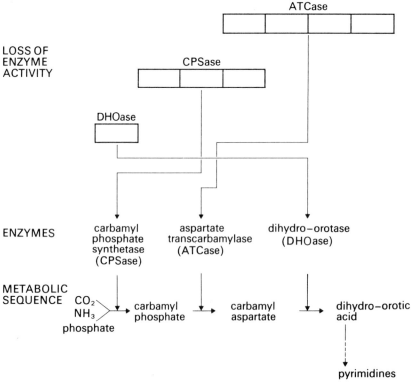

Figure 21 The relationship between the genetic fine-structure of the rudimentary gene and the enzymes it codes for, and the metabolic role of the enzymes in *D. melanogaster*.

Do not concern yourself with the biochemical detail, but rather note that here is another example in eukaryotes in which the primary molecular basis of a 'visible' mutation is now understood. This conclusion has been confirmed by the results of the injection of pyrimidines into eggs laid by $r//r$ females.

QUESTION Can you explain the reason why this experiment was undertaken?

ANSWER Feeding pyrimidines to $r//r$ females restored fertility to their eggs, and the question asked of the injection experiment was whether the eggs were infertile merely because of a lack of pyrimidines. The result of these injections into eggs, which was that fertility was restored to the eggs, confirmed this.

The role of the rudimentary-wing gene in pyrimidine biosynthesis does not by itself explain the wing phenotype (Fig. 2 on p. 240) associated with this mutation. From what we have said, it is quite conceivable that the wing effect might be due to an adjacent gene affected by mutation at the same time as the rudimentary. The following experiment shows more clearly that the wing effect is related to the biosynthesis of pyrimidines. The compound 6-azauracil is an analogue of the pyrimidines found naturally, and it specifically inhibits the final enzyme in the biosynthesis of pyrimidines, orotidylate decarboxylase. When 6-azauracil is fed to wild-type *D. melanogaster* during their development, the flies emerge with wings identical to those of rudimentary mutants! By breeding from these flies it is clear that they are not new mutations of rudimentary wing, but they are 'mimics' of the mutant caused by interfering with the same pathway affected by the mutation. An 'environmentally caused mimic' of a known mutant phenotype is called a *phenocopy*. In this instance, the rudimentary phenocopy shows that pyrimidine metabolism is certainly involved in the wing defect. There is no clear explanation of precisely why the wing should be deformed as a result of this metabolic upset, but it is suggested that the significance may be in the use of uridine (a pyrimidine) in the formation of chitin, the common structural protein component in the insect exoskeleton and appendages.

phenocopy

> **ITQ 8** Would it be correct to say that the female sterility and the change in wing shape produced in $r//r$ females are pleiotropic effects of the same mutation?

8.6.3 Induction of puffs by a steroid hormone in *D. melanogaster*

The following example provides evidence on how the specific temporal sequence of gene activity may be produced and controlled. It comes from an examination of the changes in the pattern of puffing that the insect hormone ecdysone will produce in polytene chromosomes of the larval salivary glands. During metamorphosis in the insect pupa, the imaginal discs (see Section 8.4) evert (turn inside-out) and cell differentiation occurs in the individual cells to form the structures characteristic of the products of these discs (e.g. the legs, wings and eyes). It is known that these processes of *eversion* and of cell differentiation are induced in the presence of the insect hormone ecdysone. We learned, in Section 8.3.2, that puffing is evidence of gene transcription in polytene chromosomes. If the changes of metamorphosis are brought on by gene activity controlled by the hormone, then is it possible to see the induction of these changes in genetic activity in polytene chromosomes when they are exposed to ecdysone?

eversion

Figure 22 shows a series of six photomicrographs of the same region of chromosome II from the polytene chromosomes of the salivary gland of the larva. Identical chromomeres in the six pictures are joined by lines. Figure 22(a) is the normal puffing pattern of these chromosomes before the formation of the pupa. (b)–(f) show the changes in the pattern of puffing that are seen in this region of the chromosome if the salivary gland is placed in a solution containing the hormone ecdysone. After 1 hour of incubation in this solution, two new large puffs are visible, at chromomeres *23E* and *22B* (Fig. 22(b)). After 4 hours the puff previously present at *25AC* has regressed. By 8 hours (Fig. 22(d)) the puff at chromomere *22B* has regressed again. The hormone ecdysone is thus inducing the formation of some puffs and causing others to regress. The sequences of puffing induced by ecdysone in chromosomes taken from late larval stages mirror the appearance of these same chromosomes at later stages in development. If we look back to Figure 8 on p. 351 we notice there is a peak of puffing activity at about 120 hours, at the formation of the prepupa.

It thus seems that ecdysone may be the natural 'trigger' or controlling substance mediating and integrating sequences of changes in genetic activity during this period of development.

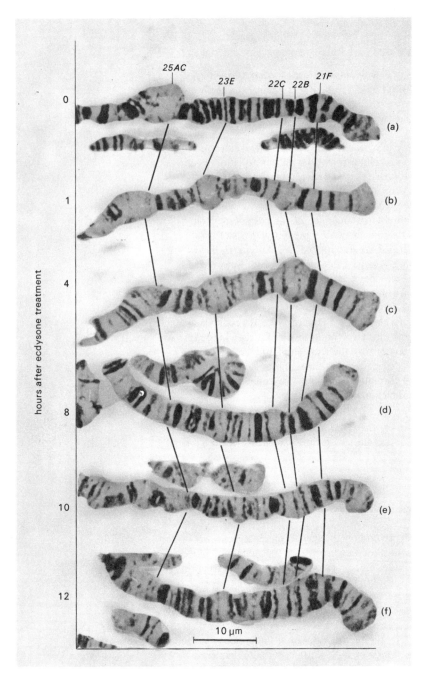

Figure 22 Photomicrographs of the same region of a polytene chromosome II in *D. melanogaster* showing the response to exposure to the hormone ecdysone, (a) normal appearance, (b) 1 hour after exposure to ecdysone, (c) 4 hours, (d) 8 hours, (e) 10 hours, and (f) 12 hours after exposure. (retouched)

8.6.4 Conclusions

It may be very obvious to say that development is a goal-directed process, that despite suffering disturbances along the way and small differences in environmental conditions, the end-product from a zygote in a particular species is usually much the same. Very seldom do adult *D. melanogaster* flies emerge with other than a 'typical' organization—two eyes, six legs, two wings of more or less constant size, and so on. This strongly implies that the process of development is *homeostatic*; that is, individual processes and structures that appear during development are interrelated and interdependent, that some are contingent upon earlier events, and that the system can take account of and correct deviations from a 'norm' of development. We do not intend to give examples to support this notion, but in this Section we have given a glimpse of the interrelations that can exist, namely:

homeostatic

1   How one event during development (in our example, the production of the insect hormone ecdysone) can 'trigger' a number of other events (puffs in chromomeres) that are, therefore, dependent upon that initial common stimulus.

2   How the basic and common metabolic interconversions (the biosynthesis of pyrimidines in our example) may be intimately related to specific developmental phenomena (the formation of the insect wing).

3   How a single metabolic defect (the failure to maintain the normal copper concentration in the blood plasma of a mouse) can result in a wide spectrum of

370

diverse phenotypic changes. It is instructive to compare the consequences of a defect like that of the mottled gene in *M. musculus* with another mutation in the same species, the 'testicular feminization syndrome' (*Tfm*), which causes another spectrum of phenotypic changes. Males carrying this mutation show a female external phenotype in spite of having a testis and an XY chromosomal constitution. It has been suggested that the effect is the consequence of a mutation in a gene that regulates the response of tissues to hormones from the testis.

The difference between the two mutations is that pleiotropic effects of *Mo*br arise from the consequences of a common defect in copper metabolism in different tissues, whereas the phenotype of *Tfm* indicates the defects that can arise when part of the network of integrating 'signals', a particular hormone in this case, fails in its determining role. Although the roles of these two genes might be considered very different, the consequences of mutations in these two genes are surprisingly similar—a collection of rather severe pleiotropic effects.

If we view development as an integrated process, it is not surprising, therefore, that certain critical genes will produce bizarre and widespread effects when they mutate. The analysis and unravelling of the consequences of changes in these genes represents an approach towards a full account or 'explanation' of the process of development.

8.6.5 Summary of Section 8.6

1 The phenotypic consequences of a mutation in the mottled gene or the *Tfm* gene in *M. musculus* or in the rudimentary-wing gene in *D. melanogaster* indicate that some mutations can exert manifold effects at the gross phenotypic level.

2 The manifold and apparently unrelated phenotypic consequences caused by one mutation are called pleiotropic effects. The phenomenon is called pleiotropy.

3 Genes that 'code for' the enzymes of basic metabolic processes can also exert effects upon development. Thus metabolic and developmental processes must be interrelated.

4 Particular (and often extreme) environmental stimuli can cause changes that mimic the effects of individual mutations. This phenomenon is called phenocopying and the environmentally produced change is termed a phenocopy.

5 A single stimulus, the hormone ecdysone, can induce transcription from a number of different genetic functional units (chromomeres). *In vitro* application of ecdysone to polytene chromosomes can change the pattern of puffing activity in a way parallel to that seen *in situ* over a period of development.

6 Mutations similar to those in man, but studied in an experimental organism like *M. musculus*, can offer enormous technical advantages for the eventual genetic analysis of the human defect.

7 The pleiotropic effects of certain mutations may be understood when development is viewed as an integrated and interdependent series of processes mediated by the activity of genes.

8.7 Attempts to dissect the genetic programme for development

By now you should be more familiar with the argument, mentioned in the Introduction, that the collection of genes in the genome of an organism may be thought of as representing a programme of instructions for development. This argument stems quite straightforwardly from two sorts of observation:

1 Individuals of each species reproduce their own kind. Even in the same environment, eggs laid by individuals of different species still develop into recognizably distinct individuals of their own species.

2 Inherited differences (mutations) may affect the process of development and the resultant phenotype.

In short, the process of development is thought of as the concerted and integrated expression of the genes. You have met an analogous situation in TV programme 6.

The process of bacteriophage reproduction and morphogenesis was dissected by studying a variety of different mutations, all of which prevented the production of complete bacteriophage progeny after the infection of a bacterium by the mutation. In that situation it was possible to define the general role of particular genes (affecting the production of 'heads', 'tails' or 'tail fibres'). In addition, there were some genes that were essential for phage morphogenesis but whose products were not built into the resultant phage particles; rather, these genes played some catalytic or inter-mediary function during assembly. In Unit 6, another class of genes was also intro-duced to you; these were important in the control of the expression of other genes.

QUESTION What were these genes? Were they frame-shift mutations, re-pressor genes, pleiotropic mutations or deletion mutations?

ANSWER They were repressor genes. These were genes in which mutations produced effects very similar to mutations in the 'structural genes' or in actual coding sequences for particular enzymes; they were shown to exert their effect by controlling the transcription of RNA from the structural genes through the agency of repressor molecules.

Repressor genes have now been clearly defined and isolated in several bacterial and viral systems. It is an attractive possibility to consider that such genes might exist in eukaryotes as 'control elements' of the imagined 'genetic programme', being concerned with the control of the activity of other genes.

Let us now consider what information there is on the nature and complexity of this so-called genetic programme, in which individual genes may be considered as the elements or 'words' from which the instructional programme is put together.

ITQ 9 What genetic procedure would you use to define the number of these essential elements (genes) in a particular programme (genome)?

The minimum estimate of this number gives the number of absolutely essential genes, but it is still possible that genes in which mutations cause subtle phenotypic changes may be overlooked. In *D. melanogaster* there are approximately 5 600 chromomeres, each believed to be carrying a genetic functional unit. Not all of these directly affect development, but the majority are defined by the developmental changes they cause when mutated. The most straightforward interpretation is that there are several thousand 'elements' in the developmental programme in *D. melanogaster*. By itself this does not tell us whether the programme is complex or simple as no one yet knows the way in which these elements are co-ordinated.

What is the role of each individual gene during development? This is rather more tricky, but it is crucial to the argument that the genome contains the programme for development. The gene is originally recognized when a mutation in that gene has been discovered. Remember that this was the classical definition of a gene (a segre-gating heritable unit; see Unit 2, Section 2.8). After comparing the mutant and wild-type phenotypes we can provide a description of the changes that accompany the introduction of the mutant into the genome. Indeed, the gene is usually named after the mutant phenotype (for example, rudimentary wing, bicaudal, bithorax). So, can we now say what the normal, wild-type copy of that gene does? Yes, it seems reason-able to say that the wild-type gene prevents the formation of the abnormality seen in the mutant. (You may not think that this has taken us very far!). Recall the gene described in phage T4 in TV programme 6 that when mutated, prevented the tail and head from assembling; to that gene was attributed a role in the assembly process. When we have been fortunate enough to find that the phenotypic change examined is at the biochemical level, it has been possible to identify the protein that is the immediate product of transcription and translation. The 'rosy' gene in *D. melano-gaster* codes for the enzyme xanthine dehydrogenase, and the rudimentary gene was found to code for three enzymes conserved in the biosynthesis of pyrimidines (Section 8.6.2). But suppose that the mutant phenotype is 'no-wings', which you met in TV programme 2, then what can one conclude about the role of this gene? It is clearly concerned in wing formation and its normal function is essential there. Further examination of the differences between normal and mutant individuals must eventually prompt experiments that give a clue to the basic and prime action of the gene. Recall the steps in the deduction of the primary effect of the rudimentary-wing gene in *D. melanogaster* or for the mottled gene in *M. musculus* (Section 8.6).

It is, therefore, quite possible to identify the area of development in which a gene is involved, though when the phenotypic change is far removed from the primary product of the gene it is often difficult to state precisely the contribution that the normal allele of the gene is making to development.

We have mentioned several other things about the 'developmental programme' in this Unit. By examining the phenocritical periods of lethal mutations it is possible to define the time at which a gene must already have been expressed. Genetic mosaics have been used to identify the particular cells and the precise periods of development during which a gene is active. The experimental system of analysing the changes in the puffing activity of individual chromomeres in polytene chromosomes of diptera was introduced in Sections 8.3 and 8.6. It is now becoming possible with this system to dissect individual sequences of genetic activity in which puffing of one chromomere is an essential prerequisite for puffing activity at a second site. It ought to be possible to gauge the complexity of a particular developmental process (e.g. wing morphogenesis in a fly) by discovering how many different (non-complementing) genes affect it. Nevertheless, there may be a catch here in that some genes can cause abnormalities in the wing because of some pleiotropic effect (e.g. rudimentary wing, Section 8.6), so that to take the number of 'wing mutants' as correct may be to overestimate the number of genes critically concerned, particularly with this process. So we find ourselves with the same picture we arrived at earlier, that the entire developmental process is integrated and interrelated so that one gene defect can cause an extraordinary array of changes that are superficially unconnected. In addition, the homeostatic nature of development may mean that when there is a defective gene in the programme, the ensuing phenotypic changes may be the endproduct of attempts by the system to minimize the disturbance. In this respect, dissection of the programme is quite unlike the idea of removing a piece (gene) from a jigsaw and then examining the resultant picture. It is rather as if the gap caused by the missing piece somehow closed up and instead created some upset or distortion in the picture!

The operon concept of the control of gene activity in prokaryotes arose as a result of the genetic analysis of mutations affecting enzyme induction (Unit 6, Section 6.5). Definite interrelationships between coding units for enzymes (structural genes) and other genetic controlling elements (promoters, operators, repressors) have been defined. This concept has been transferred to developmental analysis in eukaryotes and has led to the search for 'control genes' important in the expression and co-ordination of the other genes in the programme. Let us look to see where this leads us. What might the properties of such 'control' genes be? Mutations in them might reasonably be expected to have drastic and far-reaching consequences for the organism. Surely, then, the copper defect in the mouse (Section 8.6) fits the bill? But those pleiotropic effects arose only from the involvement of copper in enzymes in various structures and processes. There are mutations in the mouse, such as the testicular feminization syndrome (*Tfm*) and in *D. melanogaster* ('intersex', 'double sex' and 'transformer') that reverse the phenotypic sex of the individual. But are these loci so special? Any single set of sequentially expressed genes would then qualify as a 'control' element if each depended on the previous functioning gene in the sequence for its own initiation. If we take the argument to extremes, a lethal mutation would qualify, because the whole survival and development of the organism rests upon the normal function of this gene! The homoeotic mutations in *D. melanogaster* probably also come to mind when we mention 'control' genes. Figure 23 shows the effect of two

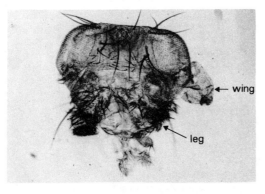

Figure 23 Photomicrograph of the head of *D. melanogaster* carrying the homoeotic mutations opthalmoptera and nasobemia.

other homoeotic mutations in *D. melanogaster*, 'opthalmoptera' (literally 'eye-wing'), and 'antennapedia'. Part of the large compound eye has been transformed into wing tissue and the base of the antenna is now a leg.

We have seen that they have several possible interpretations and it is not yet clear whether they are concerned with the maintenance of a determined state or the realization of differentiated structures in a particular imaginal disc (Section 8.4).

There is, therefore, some value in considering the idea of the genome as the programme for development because it forces us to clarify our ideas about the role of individual genes and their relative importance. There are several kinds of interrelationships that may exist between the activity of different genes during development, which may either be understood as pleiotropic effects, for example, the effect of the mottled gene in the mouse, or as genes that clearly have arisen to co-ordinate the activity of other genes.

8.8 Cautions and conclusions

Two important points are inescapable in this Unit on genetics and development. First, there has been insufficient space or time to examine the whole subject of development and so we intentionally restricted ourselves to the evidence for the involvement of genes in development and to the application of genetics as a tool or method for the dissection and analysis of development. This strategy will no doubt have led to many questions in your mind, which are still unanswered. In Section 8.8.1 a very brief account is given of approaches to development that were not mentioned earlier in the Unit! This ought to place into a clearer perspective any further reading you do. The second point is that there is no avoiding the fact that research into development is still in full spate and the subject is as yet far from being fully understood. So you can experience the excitement as well as the shortcomings of learning about a subject very much in flux. Even so, there are definite conclusions among the speculations (Section 8.8.2).

8.8.1 Other approaches to the analysis of development

Genetics is only one of a number of approaches being made to the analysis of development. Because genes, chromosomes and gene expression all lie at the heart of the developmental process it is difficult, if not unwise, to try to draw the borderline between what approach is or is not genetic. However, we shall describe a number of other approaches, all of which are contributing to our understanding of development.

1   Not only polytene chromosomes in diptera, but also all other chromosomes exhibit changes as the cells containing them change their patterns of genetic activity. Eukaryotic chromosomes contain proteins as well as DNA (the combination is often called chromatin), and important regulatory and protective functions have been attributed to various of these proteins. Chromatin can be used as a template for RNA synthesis *in vitro*. Comparisons of the RNA molecules made by cells from different tissues, or from chromatin from which various proteins have been removed, show that important differences in the transcription of genes are occurring in different cells, and these changes may be determined by the particular proteins bound to the DNA.

2   A mutated gene may be looked upon as a fine submicroscopic surgical alteration to part of the system (part of the 'programme') that allows us to see what happens. Other more familiar gross surgical operations can also be informative. By cutting or displacing parts of an embryo, one can examine the regulative properties of the developmental system. In this sense, eggs of different species vary in their regulative properties; some are assemblages of localized areas, the fates of which are already determined and other eggs may be able to regulate so as to form a complete individual from only part of an egg.

3   There is an area of developmental biology devoted to the study of morphogenesis and pattern-formation in which both the language and the level at which phenotypes are analysed differ from those of biochemistry or genetics. The principle concern is to learn about the properties of the system that result in the generation of accurate and repeated patterns of local cell-differentiation (e.g. the patterns seen in insect wings or the basic body-segmentation visible in many animals).

4   There is a multiplicity of environmental stimuli that trigger particular processes mediated by gene expression. Two examples will illustrate this. Under a given stimulus a dormant seed will germinate and cell division and differentiation will ensue. It has long been known that during animal development certain cell types will cause changes in other groups of cells as a result of interactions between the cells in what is called embryonic induction. Obviously, such phenomena offer opportunities to examine the nature of these interactions and 'triggers' through the application of biochemical methods.

5   A double-stranded DNA molecule has two complementary strands that will pair up to form a very stable association. Study of the rate at which this process of pairing of complementary strands occurs in DNA from different species has suggested that some sequences of bases in the DNA are repeated many times per genome. The existence of repeated sequences in the DNA has prompted a number of models that attempt to explain the phenomenon and also relate it to the genetic organization of the chromosomes. Remember that if a genome is a collection of single copies of many different genes, then one would not expect much, if any, repetition in DNA sequences!

This is only a glimpse, but it shows that in this Unit we have focussed on one facet of a very large and diverse field of inquiry.

8.8.2   Conclusions

We attempt below to put together in a simplified form the various aspects of development mentioned in this Unit.

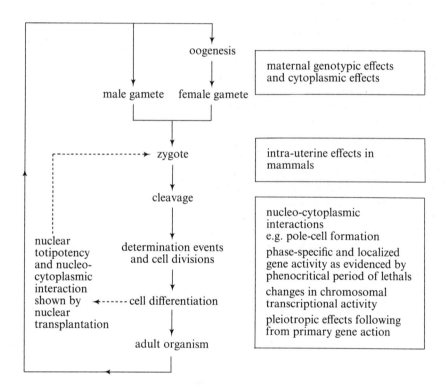

You were warned earlier that development is not yet fully understood, so you cannot expect all the answers in this Section, but there are a number of clearly established principles that have been introduced in this Unit.

The idea of the genome as a developmental programme has been sustained in this Unit. It would, however, be a mistake to see the role of the cytoplasm, the milieu of the nucleus and the genome, as a passive instrument through which the genome is expressed. Evidence has been given in this Unit that complex structural organizations can be communicated by variations in the cytoplasm. For instance, one conclusion to be drawn is that oogenesis is the end-point of a series of developmental events in a female that culminate in the often highly complex behaviour of cells surrounding the oocyte, the meiotic product that matures to become the unfertilized egg. Into that egg a second haploid genome is introduced at fertilization and the whole process then repeats itself! The cyclical nature of this process must be apparent;

in addition, it is reciprocal. That is, as a result of nuclear genetic activity, changes are created in the cytoplasm; certain of these changes are transmitted to other nuclei (see, for example, the phenomenon of the poleplasm in Section 8.5). The cytoplasm of an egg, and the cytoplasm of a differentiated cell, offer different biochemical environments to genetically identical nuclei and result in different patterns of gene expression (Section 8.2). This gene activity may function to sustain a particular physiological state in a cell over many cell divisions (e.g. the perpetuation of determined 'states' in insect imaginal discs—Section 8.4). One is forced to conclude that neither the cytoplasm nor the genes should be separately identified as the more important component because neither functions correctly without the other.

The phenomenon of cortical heredity was introduced to you in TV programme 7. In the unicellular *Paramecium aurelia*, differences in certain important structural features of the cortex, the surface layer of this cell, are inherited stably over many cell generations in the absence of any differences in nuclear genes. In that programme, Professor Sonneborn raised the possibility that similar mechanisms may be in operation in other groups of organisms, most particularly in the eggs of multicellular animals. You should compare this concept with the more orthodox view that inherited differences between individuals must eventually have arisen from differences in nuclear genes. On this latter view, differences in the behaviour of eggs (for instance, eggs laid by bicaudal females and by wild-type females in *D. melanogaster*; see Sections 8.1 and 8.5) are thought to be ultimately explicable in terms of alterations in the amount, nature or disposition of molecules in the egg or in the properties of the surface membrane, all of which would be traceable to the activity of these genes during oogenesis.

The argument has often been raised that what is true about development in one organism, may not be true for any other, in which case the picture given in this Unit largely from work with the fruit-fly, *D. melanogaster*, and from the mouse, *M. musculus*, may be an artefact. This may be countered by the argument that the basic mechanisms would be expected to be common to all organisms because of the universal nature of the genetic code and the basic biochemical attributes displayed by all organisms. Different groups of organisms may represent evolutionary developments in which particular mechanisms or 'strategies' have evolved to have greater importance in some groups than in others, yet from which the basic principles can still be discovered.

The final comment is that in this Unit we have been concerned with the relationship between genetics and phenotypic characteristics, often very complex in their manifestation. We have shown that the further removed the phenotypic characteristic is from the primary gene product (in the sense that there will have been intervening interactions with the products of many other genes and environmental influences), the more difficult and dangerous it is to ascribe clear roles to individual genes. Radio programme 8 shows the complexity of the process of wing formation in an insect. In Units 14 and 15 we shall be considering genetic variation in even more complicated phenotypes such as behaviour in man, so it is as well to be quite clear about the limitations upon the genetic analysis of development, about the current state of knowledge in this field and about the contribution of genetics to this central intriguing area of biology.

Answers to ITQs

ITQ 1 (*Objective 2*) Statements (ii) and (iii) are correct.

Statement (i) cannot be true because, demonstrably, the same nucleus subsequently went on to form a complete range of tadpole cell types, including, for example, reticulocytes containing haemoglobin, which the skin cell does not contain.

Statement (ii) is correct; this is exactly what the nucleus did.

Statement (iii) Transplanting a nucleus from one cell to another is not expected to change the genes contained in it unless the actual operation damages the nucleus. So the changed behaviour—the formation of a complete tadpole rather than just a collection of skin cells—must reflect a new influence that the nucleus came under in the egg cell.

Statement (iv) presupposes that a differentiated cell is defined as one that cannot change or become redifferentiated in other ways. This is unwarranted and patently at variance with the experimental observations.

ITQ 2 (*Objective 3*) Interpretation (iii) is correct.

The mating (a) in Table 1 yields three genotypes in the ratio $1:2:1$ in the order shown, but one of these genotypes, $l_1//l_1$, is a homozygous lethal and will die at some point, and so reduce the survival rate of the progeny of this mating to 75 per cent. There is no genetic reason to expect any losses in the progeny of mating (b) and the same random and accidental events will affect both (a) and (b) equally, so the point during development at which l_1/l_1 dies will be the point *after* which only 75 per cent of the zygotes survive. The lethal did not die in the embryo because the hatching rates from the egg in matings (a) and (b) are not significantly different ($\chi^2_1 = 0.4$, $p > 0.90$), and they are high. However, the ratio of surviving adults to those that did not survive in matings (a) and (b) is significantly different ($\chi^2_1 = 88.4$, $p < 0.001$). Only $373/500 \times 100$ per cent or 74.6 per cent of eggs in mating (a) survived to be adults. So, we must conclude that the l_1/l_1 homozygotes were 'lost' somewhere after hatching from the egg, but before emergence from the pupa to the adult.

ITQ 3 (*Objective 3*) At first glance the data seem good evidence for our basic proposition or hypothesis (see the first paragraph of Section 8.3); but statements (i)–(v) may have unsettled the picture.

Statement (i) is reasonable. The gene must have been expressed, otherwise the embryo would not have been abnormal and no phenocritical period could have been defined.

Statement (ii) is rather extreme. The first time we can see any change from the normal will be determined by our techniques of observation and comparison; for example, it will depend on whether we use a microscope, or dissect or section the material, and so on. Only if we were fortunate enough to be able to detect and identify the immediate gene product could we justify making this claim.

Statement (iii) is acceptable but somewhat conservative. All genes *may* be expressed all the time but at certain times their effects may be drastic or limiting. We would really like to know if all genes are transcribed and translated continuously in all cells. The weight of evidence is against this view which, if it were true, would knock our hypothesis for six!

Statement (iv) is only partially true and contains a *non sequitur*. It is true that the phenocritical periods of different mutants are distributed throughout embryogenesis, but each mutant has a definite phenocritical period, and in repeated determinations of this period it does not vary; it is not random at all. Therefore, the rest of the statement does not follow logically.

Statement (v) is not a valid criticism. Despite being a selected group of all possible lethal mutations, this group, nevertheless, shows that different mutations cause abnormalities at different points during embryogenesis. Even so, the substance of statement (iii) still applies to statement (v)!

ITQ 4 (*Objectives 3, 4 and 8*) (i) and (ii) are true, (iii) is not true and (iv) and (v) are possibly true, but as yet unresolved.

(i) If the gene bithorax had not been cell-autonomous in its expression, the yellow and multiple wing-hair patches would still have had the appearance of haltere tissue.

(ii) Changing the genotype in the haltere imaginal disc has caused the transformation, so the gene must have been active and must have exerted its effect in the haltere tissue. The wing imaginal disc is considerably larger than the haltere disc and this is reflected in the larger size of the wing compared with the haltere. There is some evidence that, when this transformation from haltere to wing occurs, there is compensatory extra growth in the 'wing' tissue so that the bithorax gene, in addition to affecting the final cell-differentiation process, is presumed to be active during the growth of the imaginal discs and cell division.

(iii) If the bithorax gene were involved only in an initial determination event in the embryo, clones formed late in development would change the genotype when this event had already occurred, and would not be expected to cause any transformation, and yet they do.

(iv) and (v) These are still unresolved alternative explanations, although if (iv) were true we would need to explain why this lag or 'dilution' effect did not apply to the genes yellow body or multiple wing-hair. A distinction between (iv) and (v) might be made if the clones could be caused to undergo further cell divisions before differentiation. In which case, if (iv) were the explanation, the wing transformation might appear, but further divisions ought not to change the phenotype according to explanation (v).

ITQ 5 (*Objectives 4 and 8*)   (a) (i), (b) (ii), (c) (iii).

(a) The zygote starts with a nucleus carrying two X-chromosomes. It loses one (ring-shaped) X-chromosome in one of the first two division products so producing an XX and an XO nucleus. Each of these subsequently divides normally without further loss so that 50 per cent of the cells will be XX and 50 per cent will be XO in karyotype. Phenotypic sex differences in *D. melanogaster* are determined by the number of X-chromosomes in relation to the other chromosomes, and are cell-autonomous so that cells containing XX and XO karyotypes can be next to each other and yet form contrasting structures. There is no Y-chromosome in the original zygote.

(b) The axis of the first mitosis is random and the products, the XX and XO nuclei, can occupy any position; thereafter no further loss of chromosomes occurs and the division products of each of the two nuclear types tend to stay together, giving rise to a clear border between XX and XO tissue in the adult. There is good evidence from the fate-mapping and focus-mapping procedures in Benzer's paper against the occurrence of much cell migration or shuffling.

(c) The normal rod-shaped X-chromosome carries the recessive allele of the yellow-body gene (*y*) together with the recessive mutant allele of the developmental mutation being analysed. The ring-shaped X-chromosome has the dominant, wild-type alleles of both genes, so XX cells will be wild-type for both characteristics. Only the XO cells carry the mutant alleles and are recognizable because of the pigmentation change they produce. It is in these areas that one looks for signs of the effect of the mutation. If the XO cells in any gynander include the focus of the developmental mutation, and it is cell-autonomous in effect, then those gynanders will show the abnormality.

ITQ 6 (*Objective 2*)   The concept of totipotency would predict that the somatic cell nuclei, with only 8 chromosomes, ought to be able to assume the functional role of the zygotic nucleus, which has 38 chromosomes. If all the chromosomes are carrying information, then totipotency cannot apply in this species. Because of the curious behaviour of chromosomes in the first few divisions, this species remains as an exception to the rule of totipotency.

ITQ 7 (*Objectives 3 and 5*)

(a)
$$\frac{fs}{fs} \quad \text{and} \quad \frac{fs}{\underline{\hspace{1cm}}}$$

(b)
$$\frac{fs}{fs^+} \quad \text{and} \quad \frac{fs}{\text{www}\bullet}$$

(c)
$$\frac{fs}{fs} \quad \text{and} \quad \frac{fs}{fs^+}$$

$$\frac{fs}{\text{www}\bullet} \quad \text{and} \quad \frac{fs^+}{\text{www}\bullet}$$

(i) Incorrect. If the statement were true then all the parents except the wild-type male in cross (b) would be dead.

(ii) True. This is the common denominator between crosses (a) and (b). Contrast this with the genotype of the female parent in (c).

(iii) Incorrect. The *fs/ /fs* female and *fs/Y* male parents would never have lived to mate.

(iv) Incorrect. All the female progeny in cross (b) are formed by fertilization with sperm carrying fs^+, but they die too.

(v) Incorrect. We were able to make crosses (a) and (b) and so these genotypes must be able to survive under some conditions.

ITQ 8 (*Objective 7*) Yes. Both phenotypic effects are produced in the *r/ /r* female genotype, the one affecting her eggs and the other her wings, and we described how deficiencies in the biosynthesis of pyrimidines underlie both effects. Compare this situation with the example of pleiotropy given in Section 8.6.1.

ITQ 9 (*Objective 8*) The question is to find out how many genes in the genome are necessary for normal development. This can only be done genetically by seeing how many different genetic functional units can be defined that, when mutated, affect development. 'Different' will be defined by a complementation test (Unit 2, Section 2.2 and Unit 5, Section 5.5).

Self-assessment questions

Section 8.3

SAQ 1 A new temperature-sensitive lethal mutation has been discovered in *D. melanogaster* and is being grown as a homozygous stock at 22 °C, at which temperature the flies live. When eggs from this stock are grown at 29 °C, no adult flies are produced. A series of identical samples of eggs from this stock laid at 22 °C are put in food vials and subjected to a shift in temperature (either 22 °C or 29 °C) at particular periods during development. For example, one sample is cultured at 22 °C for the first 12 hours and then is transferred to 29 °C for the rest of its development. Table 3 shows the outcome of a series of such temperature-shift experiments.

Table 3 Results of temperature-shift experiments with a temperature-sensitive mutation in *D. melanogaster*

Time of shift from 22 °C–29 °C (in hours)	Result	Time of shift from 29 °C–22 °C (in hours)	Result
12	−	12	+
24	+	24	−
36	+	36	−
48	+	48	−
60	+	60	−
72	+	72	−

+ = adult flies emerged
− = no adult flies produced

Deduce from these results when the *temperature-effective period* is for this gene, that is, the period of growth when the higher (non-permissive) temperature kills the developing flies, and suggest how you would confirm it by a 'double-shift' experiment, that is, by two successive shifts during development.

temperature-effective period

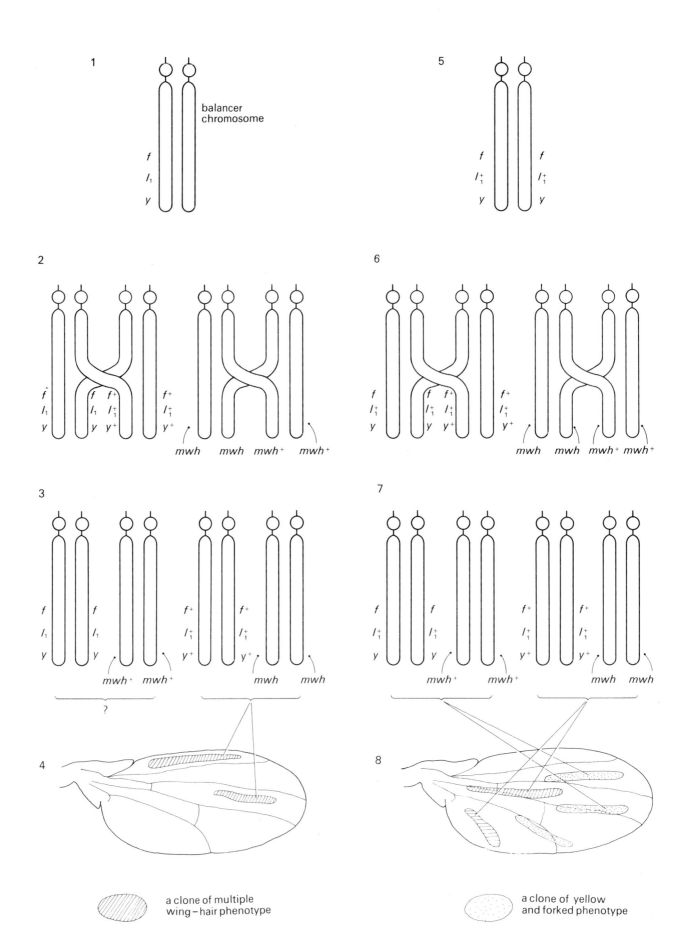

Figure 24 The steps in forming a genetic mosaic for a recessive lethal mutation in *D. melanogaster*: stage 1, the genotype of the heterozygote for the lethal mutation; stage 2, the formation of a heterozygote carrying the lethal and three other mutations; stage 3, the induction of mitotic crossing over in the X-chromosome and IIIrd chromosome; stage 4, the appearance of a typical wing showing the mosaics; stages 5–8, the same steps for the genotype not carrying the lethal mutation.

Section 8.4

SAQ 2 A recessive lethal mutation (l_1) is discovered in the X-chromosome of *D. melanogaster*. By recombination, it is combined with the recessive, cell-autonomous, mutations, yellow body (y) and 'forked bristle'(f), and maintained using a balancer chromosome (see Unit 5, Section 5.3), as shown in Figure 24, stage 1. Heterozygotes, between this genotype and another stock carrying the recessive cell-marker mutation, multiple wing-hair (*mwh*), in chromosome III, are made and subjected to X-irradiation during development to induce mitotic crossing over in some cells (Fig. 24, stage 2). In a parallel experiment, another heterozygote is formed (Fig. 24, stage 5), which differs from that in stage 1 in that it does not contain the lethal l_1 and so lives as a homozygote. This second heterozygote is also X-irradiated during development to induce mitotic crossing over in some cells (Fig. 24, stage 6). If mitotic crossing over occurs in the X-chromosome, one of the two products will be homozygous for yellow body and forked bristle and a patch of cells of that phenotype will be seen. Similarly, if mitotic crossing over occurs in chromosome III then one of the cells will grow to form a patch showing the multiple wing-hair phenotype.

Figure 24 (stages 4 and 8) shows a representative wing from an adult of each genotype after irradiation. Notice that although both types of patch are found in stage 8, only patches of multiple wing-hair phenotype are seen in stage 4.

Select the correct statement(s) from 1–5 below.

1   Mitotic crossing over does not occur in genotypes carrying the lethal mutation l_1.

2   Mutation to yellow body (y) or to forked bristle (f) is less effective in the presence of the heterozygous lethal l_1.

3   The gene in which mutation l_1 lies is not expressed in the wing during development.

4   Patches of yellow and forked-bristle phenotype are not seen in stage 4 because l_1 is a cell-autonomous lethal mutation.

5   The gene in which l_1 lies is a possible 'control' mutation because it affects the expression of the other linked genes, yellow body and forked bristle.

Section 8.5

SAQ 3 In *D. melanogaster*, after the series of rapid nuclear divisions that occurs early on in the centre of the fertilized egg, the first nuclei to be included within cell membranes are those that form the pole cells in the posterior tip of the egg. Figure 25 is a view of the egg of *D. melanogaster* showing the pole cells. In the rest of the egg, the nuclei are still in a common cytoplasm.

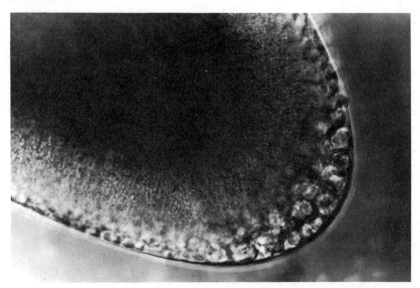

Figure 25 Photomicrograph of part of the egg of *D. melanogaster* at an early developmental stage in which only the pole-cell nuclei have formed cells.

These pole cells are the germ-line; they eventually give rise to the cells of the ovary or testis of the adult. The following experiment was carried out to see whether the cytoplasm (the poleplasm) in the area eventually occupied by the pole cells specifically determined nuclei entering that region to be pole cells. Cytoplasm from the pole-plasm area of one egg (of wild-type genotype) was injected into the anterior tip of another egg (of genotype, multiple wing-hair and ebony body) at the same, pre-cellular, stage of cmbryogenesis (Fig. 26, stages 1 and 2).

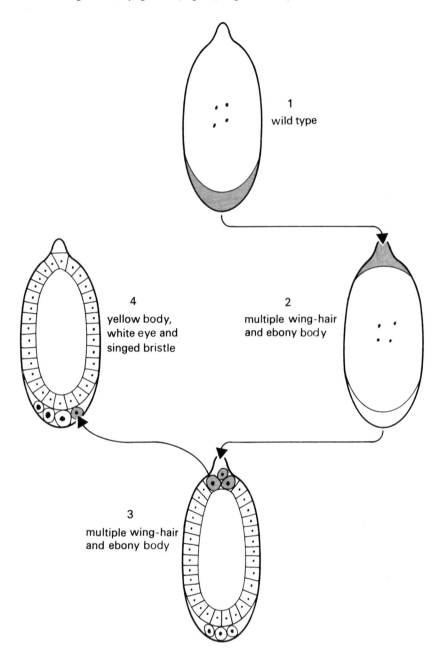

Figure 26 The effect of transplanting poleplasm in *D. melanogaster* to the anterior of an egg at an early stage: stage 1, the removal of the poleplasm; stage 2, the injection of poleplasm into the anterior of another egg; stage 3, the formation of pole cells in the anterior of the egg; stage 4, the transplantation of these pole cells to the posterior of a third egg, and the integration of the transplanted pole cells into the reproductive cell-line of the host.

When these injected embryos had developed further, it was clear that cells similar in appearance to pole cells had formed at the site of the injection of poleplasm (Fig. 26, stage 3). In order to discover whether these cells were functional pole cells they were transported to the pole region of an embryo of yet a third genotype, 'yellow body, white eye and singed bristle' (Fig. 26, stage 4).

When these embryos emerged as adults they were crossed to other flies of genotype, yellow body, white eye and singed bristle. All the mutations mentioned are recessive. If the transplanted cells were indeed normal pole cells then their genotype ought to be detectable in the progeny. Some of the progeny were wild-type in phenotype and not all yellow-bodied, white-eyed and singed-bristled, as they would be if yellow-bodied, white-eyed, singed-bristled flies had been crossed to each other. In addition, the wild-type progeny were shown by further crosses to be carrying the mutations multiple wing-hair and ebony body.

Select from 1–5 the statements that are reasonable interpretations or explanations of the results of the experiment.

1   Pole cells form or can be caused to form only at the posterior tip of the egg of *D. melanogaster*.

2   Poleplasm from the embryos in Figure 26, stage 1 induced the formation of pole cells in an anterior position (Fig. 26, stages 2 and 3). This is an example of a determination event mediated by a particular cytoplasmic component.

3   The appearance of the alleles, multiple wing-hair and ebony body among the progeny of flies in Figure 26, stage 4 shows that poleplasm is mutagenic at this stage.

4   Different genotypes were used for each transfer operation to guard against the possibility that in Figure 26, stage 1 nuclei might have accidentally been transferred, and in Figure 26, stage 3 to ensure that the believed pole cells formed could be distinguished after transfer (in Figure 26, stage 4) from the pole cells already present and formed by the host (of genotype, yellow body, white eye and singed bristle).

5   An anterior site was chosen for the transfer of poleplasm in Figure 26, stage 2 because pole cells are not normally formed in that region. Putting the poleplasm into the rear tip of the embryo would not have had detectable effects.

Section 8.6

SAQ 4   We show a 'pedigree of causes' of phenotypic effects seen in a rat carrying a recessive lethal mutation that results in the over-production of cartilage.

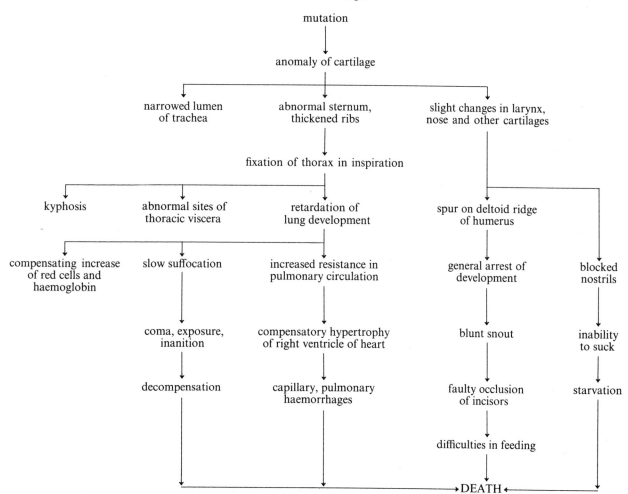

(a)   Which genetic phenomenon is this particular case an example of: position effect, phenocopy, totipotency, localized gene action, phenocritical period or pleitropy?

(b)   What aspect of the process of development does this example illustrate very clearly?

SAQ 5   Consider the following experimental observations and then answer the questions.

The puffs in specific chromomeres of polytene chromosomes of the salivary-gland cells of *D. melanogaster* that are induced by the steroid hormone ecdysone (see Section 8.6.3) can be divided into two classes, early puffs and late puffs. The early puffs appear within 15 minutes of the application of ecdysone but the late puffs appear only after a lag of at least 3 hours. The induction of late puffs is suppressed by inhibitors of protein synthesis but early puffs are unaffected.

A heat-shock of short duration delivered to salivary-gland cells induces seven specific 'heat-shock' puffs in known chromosomal locations. A diploid-cell culture derived from embryonic cells of D. melanogaster responds to the same heat-shock by synthesizing seven new species of proteins. This was discovered by comparing the total proteins produced before and after the heat-shock. Radioactively-labelled RNA synthesized in 'heat-shocked' cultured cells differs from that made in cells not exposed to heat-shock in that new different RNA molecules are made. These new species of RNA have been shown to be complementary to DNA sequences in the region of the seven heat-shock puffs.

From the data given, which of the conclusions 1–5 are reasonable?

1 The heat-shock proteins seen in the embryonic cell-line cannot be derived from the genetic sites represented by the heat-shock puffs in the polytene chromosomes because the diploid cell-line does not contain polytene chromosomes.

2 The late puffs represent the sites of recessive mutations and the early puffs indicate dominant mutations.

3 Ecdysone acts by stimulating the synthesis of proteins in cells.

4 The induction of late puffs is mediated by a step dependent upon protein synthesis.

5 The correlation between the action of cultured cells and puffing changes after the same environmental stimulus suggests that the puffing in polytene chromosomes is a reasonable indication of changes in transcriptional genetic activity in chromosomes in general.

Section 8.7

SAQ 6 Select the best definition of the terms 1–13 below from the statements A–O.

1 phenocritical period of a mutation
2 lethal phase of a mutation
3 totipotency
4 a gene with cell-autonomous expression
5 a pleiotropic mutation
6 a chimaera
7 a phenocopy
8 a homoeotic mutation
9 a gynander
10 a genetic mosaic individual
11 focus mapping of a gene
12 fate-mapping of an organ or tissue

A a mutation that produces a number of apparently unrelated phenotypic effects

B a mutation that upsets homologous pairing of chromosomes and thus produces chromosome loss during development

C the localization of the clonal precursor cells of a particular adult structure to a specific region of the egg's surface

D variation in the phenotypic effect of the same genotype dependent upon the altered place of a particular gene in the genome

E the time at which the phenotypic effect of a mutation first becomes observable as a change from the normal sequence of development

F the localization of the critical region, or the group of cells in an organism in which a mutation must be present in order to exert its phenotypic effect

G an individual that contains cells of at least two different genotypes in its somatic cells, yet is derived from one fertilized egg

H   a behavioural mutation that results in an altered response of the organism to light stimuli

I   an environmentally caused phenotypic change, closely resembling the phenotype of a known mutation

J   a gene whose phenotypic expression is limited to the cells that actually carry the particular allele involved and that produces a distinctly different cellular phenotype even when adjacent to cells of different genotype

K   the time at which a lethal mutation causes the death of the individual carrying it

L   a fly that is phenotypically part male and part female as the result of the loss of one X-chromosome from some of the clonal descendants of an original XX zygotic cell

M   a mutation that changes one structure into a distinctly different but homologous structure

N   an individual that contains cells of at least two different genotypes derived from more than one fertilized egg among its somatic cells

O   the demonstrated ability of a nucleus from a differentiated somatic cell to support the development of a complete range of normal cell types

Answers to SAQs

SAQ 1 (*Objective 3*) The temperature-effective period for this mutation is somewhere between 12 and 24 hours after egg deposition. Any transfer that was kept at 29 °C for that period produced no flies.

To confirm this conclusion one would take two samples of eggs. The first is kept at 29 °C except between 12 and 24 hours, and it should give viable flies; but the converse experiment of keeping the vial at 22 °C, except for a shift up to 29 °C between 12 and 24 hours, should kill the developing embryos. This is summarized diagrammatically in Figure 27.

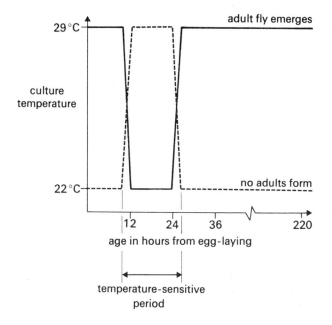

Figure 27 Temperature-shift experiments in *D. melanogaster* to confirm the location of the temperature-effective period for a temperature-sensitive mutation.

SAQ 2 (*Objectives 3 and 4*) The only correct statement is statement 4. Statement 1 is untrue because patches of multiple wing-hair phenotype do appear after X-irradiation. It is most unlikely that the patches are the result of mutation because the yellow-body and forked-bristle phenotypes always occur together, a coincidence if due to mutation. Patches of yellow and forked can be be formed by X-irradiation in the developing wing (See Fig. 24, stage 8) so it is most likely that they are also formed by the events shown in Figure 24, stage 3, but that the cells die because they are homozygous for l_1. Such cells would die only if the lethal effect were cell-autonomous and not supported by some diffusible substance from neighbouring cells

385

(statement 4). If the gene were unnecessary in the wing, a failure in that function, that is, in homozygous cells $l_1//l_1$ would not be expected to cause the cells to die (statement 3). The yellow and forked phenotypes fail to appear in Figure 24, stage 4 because l_1 causes patches of the homozygous genotype

$$\frac{y \quad l_1 \quad f}{y \quad l_1 \quad f}$$

to die; the mutation l_1 is not affecting the expression of the other genes in a developmental sense.

SAQ 3 (*Objective 5*) The reasonable statements are 2, 4 and 5. The over-all result of the experiment indicated that pole cells could be induced in an area other than the posterior tip of the egg, by transplantation of poleplasm. Thus, statement 1 is false, but statement 2 is true. Statement 3 offers insufficient evidence for us to conclude that mutation has occurred. First, only specific genotypes appeared in the progeny of the flies in Figure 26, stage 4, whereas there is strong circumstantial evidence that these alleles could have been derived via the transplantation procedure. Second, one would expect to discover evidence for frequent mutation in the germ-line in normal flies not subject to transplantation if the poleplasm were indeed mutagenic. Without the use of the genetic differences (the 'markers') between the various embryos, one could never conclude (statement 4) that the anterior transplantation had achieved anything, as pole cells are expected to form at the posterior tip of the embryo (stage 3) in Figure 26. The same objection applies to statement 5, because even if the poleplasm were not the causative agent in the determination of the germ-line, we still know that pole cells will appear in this area. So any experimental alteration in the posterior tip might be totally uninformative about the determination of pole cells.

SAQ 4 (*Objective 7*) (a) Pleiotropy. The initial disturbance to cartilage synthesis in the rib-cage leads to defective ventilation of the lungs. As a result physiological changes occur in order to compensate for the reduced efficiency of oxygen supply to the blood. Other cartilage defects affect feeding.

(b) This example illustrates the interdependence of several distinct developmental processes, even though the interactions are largely mechanical (i.e. difficulties in breathing and sucking that affect the physiological state of the young rat, causing secondary changes). It would be equally true to say that this example shows how difficult it might be to relate phenotypic changes like faulty occlusion of the incisor teeth, haemorrhages and suffocation to one primary gene defect.

SAQ 5 (*Objective 6*) Conclusions 4 and 5 are the only reasonable ones. Conclusion 1 falsely assumes that polytene chromosomes are in no way similar to normal chromosomes of diploid cells—but, of course, both chromosomes carry the same genes in the same sequence and the only difference is in the lateral multiplicity of the polytene chromosome.

Conclusion 2 is unacceptable as there is no evidence given at all that genetic dominance or recessiveness is related to the rate of transcriptional response to a stimulus.

Conclusion 3 is also unwarranted as no evidence has been provided that shows that protein synthesis in general has been altered, and indeed the early puffs can be induced by ecdysone in the presence of inhibitors of protein synthesis.

SAQ 6 (*Objective 1*)

1 E	4 J	7 I	10 G
2 K	5 A	8 M	11 F
3 O	6 N	9 L	12 C

Bibliography and references

Markert, C. L. and Ursprung, H. (1972) *Development Genetics*, Prentice-Hall.
(A very readable 200-page introductory text.)

Gurdon, J. B. (1974) *Control of Gene Expression in Animal Development*, Oxford University Press.
(A very clear exposition of experiments on nuclear transplantation and its use in investigating questions of gene expression during development.)

Hadorn, E. (1961) *Developmental Genetics and Lethal Factors*, Methuen.
(Now classical, this is a detailed account of the field by a pre-eminent experimentalist. Despite its publication date the book has hardly dated in its relevance.)

Acknowledgements

Grateful acknowledgement is made to the following for illustrations used in this Unit:

Figure 3(a), and (b) from M. Bownes, 'A photographic study of development in the living embryo of *D. melanogaster*' in *Journal of Embryological and Experimental Morphology*, **33**, 1975; and *Figures 3(c) and 25* by courtesy of M. Bownes, Biology Dept., University of Essex; *Figure 4* from J. B. Gurdon *et al.*, 'The developmental capacity of nuclei transplanted from keratinised skin cells of adult frogs' in *Journal of Embryological and Experimental Morphology*, **34**, 1975; *Figure 5* from Prof. T. R. F. Wright, *Advances in Genetics*, ed. W. Caspari, **15**, Academic Press, 1970; *Figure 6* from M. Ashburner, 'Patterns of puffing activity in the salivary gland chromosomes of *Drosophila* V' in *Chromosoma*, **31**, 1970, Springer-Verlag; *Figures 7 and 8* Ibid. I, **21**, 1967; *Figure 13* from H. Wildermuth, in *Science Progress*, **58**, 329–58, Blackwell Scientific Publications, 1970; *Figure 16* by courtesy of Dr. G. Morata, MRC Molecular Biology Lab., Cambridge; *Figure 19* by courtesy of Dr. D. M. Hunt, Queen Mary College, London; *Figure 20* by courtesy of Dr. J. A. Walker-Smith, Queen Elizabeth Hosp. for Children, London; *Figure 22* from M. Ashburner, Ibid. VI, **38**; *Figure 23* by courtesy of Dr. W. J. Ouweneel, Hubrecht Laboratory, The Netherlands; *Table* in SAQ 4 from Prof. H. Grüneberg, 'An analysis of the "pleiotropic" effect of a new mutation in the rat' in *Proc. Roy. Soc. Lond. B*, **125**, Cambridge University Press, 1938.